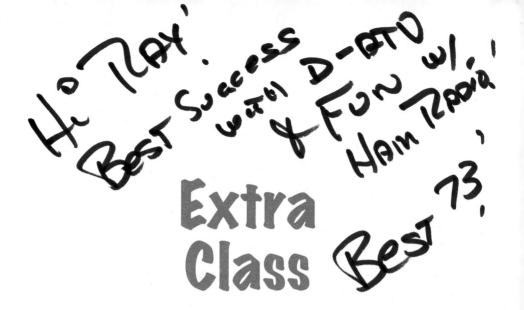

Hi Ray! Best Success with D-RTD & Fun w/ Ham Radio! Best 73,

Extra Class

FCC License Preparation
for
Element 4
Extra Class Theory

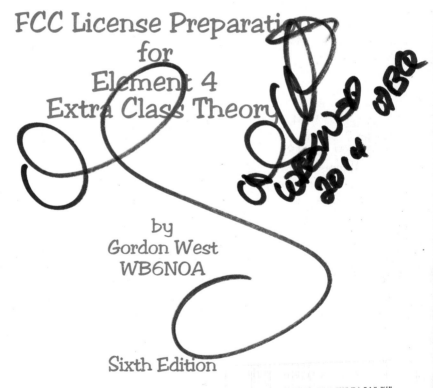

by
Gordon West
WB6NOA

Sixth Edition

Master Publishing, Inc.

This book was developed and published by:
Master Publishing, Inc.
Niles, Illinois

Editing by:
Pete Trotter, KB9SMG

Thanks to: Suzy West, N6GLF; Chip and Janet Margelli, K7JA and KL7MF;
Clint, K6LCS; Jim Ford, N6JF; Don, W6GPS; Jim and family, N3UZ; Tom,
W6WC; George, Mimi, and Rosie, WA6RIK; Marilynn Jordan, KE6RIN; Eric,
KL7AJ; and Dr. Joe, AC2FQ

Photograph Credit:
All photographs that do not have a source identification are either courtesy of
RadioShack, the author, or Master Publishing, Inc. originals.
Cover photos by Julian Frost, N3JF, and Russell Decker, KB6YAF.
Mr. Elmer and cartoons by Carson Haring, AC0BU

Printing by:
Arby Graphic Service
Niles, Illinois

Sixth Edition
  5   4   3   2   1

Table of Contents

QUESTION POOL NOMENCLATURE

The latest nomenclature changes recommended by the Volunteer Examiner Coordinator's Question Pool Committee (QPC) for the Element 4 question pool effective July 1, 2012 and valid to June 30, 2016 have been incorporated in this book.

FCC RULES, REGULATIONS AND POLICIES

The NCVEC QPC releases revised question pools on a regular cycle, and deletions as necessary. The FCC releases changes to FCC rules, regulations and policies as they are implemented. This book includes the most recent information released by the FCC when this copy was printed.

Preface

Amateur Extra Class is the highest-level Federal Communications Commission license you can hold in the amateur radio service. Achieving the Extra Class license immediately opens up exclusive sub-bands for voice operation on worldwide frequencies, along with additional, exclusive Extra Class sub-bands for Morse code and data. Your Extra Class license gives you full operating privileges across the complete spectrum of every ham radio band!

The Morse code test has been totally eliminated for all classes of amateur radio license. Prior to 2000, Extra Class applicants were required to pass a 20 word-per-minute Morse code test. From 2000 through 2006, the code test for General Class was reduced to 5 words-per-minute, which also satisfied the Extra Class code requirement. On February 23, 2007, all Morse code testing was completely eliminated by the FCC. And guess what? More hams than ever are learning the code and practicing CW over the worldwide airwaves, with no fear of a code "stress test!" So if you haven't already done so, I encourage you to learn the code. It's fun!

Achieving the Extra Class license allows you to administer all amateur radio exams. As an Extra Class licensee, you may be accredited as a Volunteer Examiner to prepare and administer ham exams as a member of a 3-person VE team. You also could qualify to become a contact volunteer examiner to help coordinate ham exam teams in your area. This is a great way to help our hobby and service grow!

The new Element 4 Extra Class question pool included in this book is valid for written examinations from July 1, 2012 through June 30, 2016. The total number of questions in the new pool is about the same as in years past – 702, with 50 of these multiple choice questions appearing on your upcoming Extra Class exam. No additional math complexity has been added to this fresh new pool, but there are many new questions on newer operating modes and techniques. My book explains everything about the Extra Class:

- *Chapter 1* reviews the operating privileges you'll earn when you pass your Extra Class exam.
- *Chapter 2* is a brief review of the history of ham radio licensing and explains the exam requirements for all of the current FCC Amateur Radio licenses.
- *Chapter 3* contains all 702 Element 4 questions and answers, along with my full description of the correct answer, followed by the correct answer key.
- *Chapter 4* tells you what to expect on examination day, and reviews the latest FCC e-filing requirements.

Welcome Advanced and General Class operators to Extra Class exam preparation! Get set for new privileges and increased VE opportunities. I hope to hear you on our top bands very soon!

Gordon West

73!

Extra Class Privileges

The Extra Class amateur operator/primary station license permits you to use all segments of every worldwide band for long-distance voice, data, and video amateur communications, expanding your General Class or grandfathered Advanced Class operating privileges. Of course, you also keep all of the privileges you earned as a General or Advanced Class operator.

Extra Class has always been recognized as the "top" amateur radio license because of the unlimited skywave privileges it provides on each of the worldwide ham bands. Earning the Extra Class privileges requires passing written exams for Element 2, Technician Class, and Element 3, General Class, prior to taking your Element 4 exam.

Never in the history of amateur radio has it been easier than now to achieve the Extra Class license. There no longer is a Morse code test required. Certainly, the old 20 wpm code requirement was a huge obstacle for many accomplished ham radio operators who wanted to achieve Extra Class but just couldn't get past that tough, tough CW test. In February, 2007, the Federal Communications Commission ended code test requirements for all amateur radio licenses, concluding that "...this change eliminates an unnecessary burden that may discourage current amateur radio operators from advancing their skills and participating more fully in the benefits of amateur radio." Most other nations had already dropped their Morse code test requirements, too.

Just because the code test has been eliminated, it doesn't mean earning your Extra Class license will be a snap. For most of us, it requires having spent time developing our operating skills, and means spending some serious time diligently studying the electronic theory and operating techniques asked in the question pool. Once you pass your exam, becoming an Extra Class means you've reached the top of our hobby and service.

The Extra Class license allows you to give something back to the amateur service as a volunteer conducting examinations. As an Extra Class amateur operator, you may become a lead Volunteer Examiner. You are qualified to prepare and administer all levels of amateur radio testing as a member of a 3-person accredited VE team. You also can help recruit and organize other hams to serve on your VE team. This is a great opportunity to help our hobby grow and prosper!

When you pass your Extra Class examination, you will gain a 25-kHz "window" of exclusive CW privileges at the bottom end of each of the 80-, 40-, 20-, and 15-meter bands. You also gain a total of 150 kHz of additional "windows" of exclusive Extra Class phone privileges on 75-, 20-, and 15-meters. If you are a General Class operator upgrading to Extra, you will gain valuable DX operating privileges on the Advanced and Extra Class bands. Remember, grandfathered Advanced Class operators continue to enjoy their own sub-band privileges indefinitely.

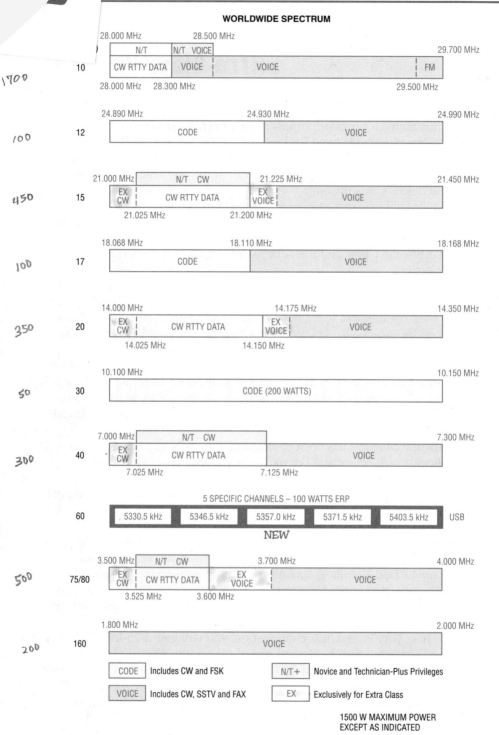

WORLDWIDE SPECTRUM

Figure 1-1. Extra Class Privileges

1500 W MAXIMUM POWER
EXCEPT AS INDICATED

CODE — Includes CW and FSK
VOICE — Includes CW, SSTV and FAX
N/T+ — Novice and Technician-Plus Privileges
EX — Exclusively for Extra Class

As an Extra Class operator, you will be eligible for the AA-ALx2 or 2x1 short-block call signs that are reserved exclusively for those amateurs who have made it to the very highest level of license in the amateur service. You may wish to take advantage of the Vanity Call Sign program, too, to select specific, available call letters of your choice (see Chapter 4 for details on this). And probably best of all, when you reach Extra Class you do not have to prepare for anymore license upgrades. You're the top!

Now, let's take a look at privileges of the Extra Class license.

THE EXTRA CLASS LICENSE

Let's begin by taking a look at *Figure 1-1*, which graphically illustrates your overall Extra Class privileges on the 160-meter medium-frequency (MF) band from 1.8 to 2.0 MHz, and on the 80-, 60-, 40-, 30-, 20-, 17-, 15-, 12-, and 10-meter high frequency (HF) bands from 3.0 to 30 MHz.

Morse code privileges – where only CW and data are allowed – are shown in the designated area on the left side of each band edge. The areas that include voice privileges are shown in the designated areas on the right. In the 80-, 20-, and 15-meter bands, we illustrate areas exclusively for Extra class that are not shared with grandfathered Advanced Class operators. If you are upgrading from Advanced Class, passing the Extra Class exam immediately gives you access to these Extra Class sub-bands.

If you're upgrading from General Class, you gain all Extra Class sub-bands, and you also gain all grandfathered Advanced Class sub-band privileges, too. In other words, as an Extra Class operator, *you get everything!*

Take a look at *Figure 1-1* for Extra Class HF license privileges and see the "EX" – this is where you have *exclusive* voice operating privileges and CW privileges shared with no one other than Extra Class operators.

EXTRA CLASS LICENSE PRIVILEGES

160 METERS: 1.8 MHZ - 2.0 MHZ.

1.800 MHz
2.000 MHz
160
VOICE

Your Extra Class privileges are the same as Advanced and General on this band. You may operate voice and code from one end of the band to the other. The 160-meter band is great for long-distance nighttime communications. It is located just above the AM broadcast radio frequencies. At night, 160 meters lets you work the world.

75/80 METERS: 3.5 MHZ - 4.0 MHZ.

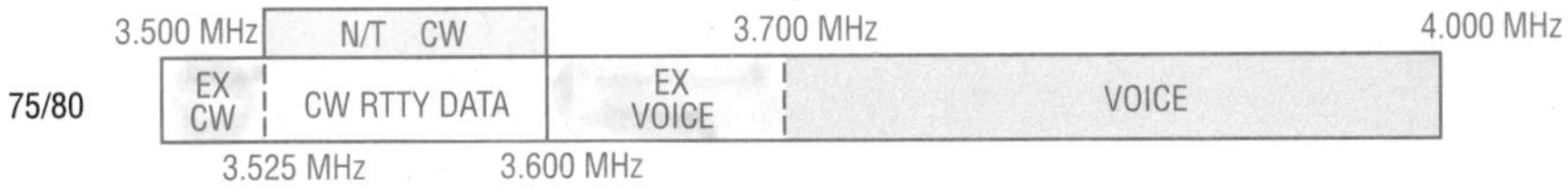
3.500 MHz
N/T CW
3.700 MHz
4.000 MHz
75/80
EX CW
CW RTTY DATA
EX VOICE
VOICE
3.525 MHz
3.600 MHz

As an Extra Class operator, your exclusive voice privileges have recently been expanded to include 3.600 to 3.700 MHz, not shared with any other class of ham license. This is a huge increase for nighttime DX voice contacts, well over 6000 miles away. You also gain exclusive CW privileges at the bottom of the band.

60 METERS: 5 SPECIFIC CHANNELS

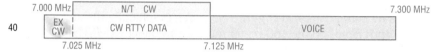

Recently, the middle channel on 60 meters has been changed from 5366.5 kHz on your dial to 5357.0 kHz, upper sideband. You now may also run up to 100 watts Effective Radiated Power, referenced to a half-wave dipole.

40 METERS: 7 MHZ - 7.3 MHZ.

Extra Class operators have an exclusive 25 kHz of CW operating at the bottom of the band. Voice privileges now begin at 7.125, a recent increase in "elbow room" on the 40 meter band. At nighttime, it's possible to operate both code as well as voice throughout the world on 40 meters. However, 40 meters is shared with worldwide AM shortwave broadcast stations, so at night be prepared to dodge megawatt carriers playing everything from rock and roll to political broadcasts. During daylight hours, the range on 40 meters is approximately 500 miles.

30 METERS: 10.1 MHZ - 10.15 MHZ.

This is a shared band for code only between General, grandfathered Advanced, and Extra Class operators. 30 meters is located just above the 10-MHz WWV time broadcasts on shortwave radio. Voice is not allowed on this band by any class of amateur operator. Maximum PEP power level is limited to 200 watts.

20 METERS: 14 MHZ - 14.350 MHZ.

Extra Class operators have 2 band segments of exclusive frequency privileges. For CW, you gain the bottom 25 kHz of the band. For voice, you have exclusive use of 14.150 to 14.175 MHz, not shared with any other class of license! Happy DX! Of course, since the Extra Class license offers all band privileges, you can operate on those shared frequencies for Advanced and General Class, too.

20 meters is where the real DX activity takes place. Almost 24 hours a day, you should be able to work stations in excess of 5,000 miles with a modest antenna set up. If you are a mariner, most maritime mobile operations take place on the 20-meter band. If you are into recreational vehicles (RVs), there are nets all over the country especially for you. And if you enjoy DXing the world and collecting rare QSL cards, 20 meters is where the action is!

17 METERS: 18.068 MHZ - 18.168 MHZ.

	18.068 MHz	18.110 MHz	18.168 MHz
17	CODE	VOICE	

There is plenty of elbow room here with a lot of foreign DX coming in during the day and evenings. While the Extra Class operator does not have any exclusive portions on 17 meters, Extra Class licensees have the entire band from 18.068 MHz all the way up to 18.168 MHz. Voice begins at 18.110 MHz, and is shared with General and grandfathered Advanced operators.

15 METERS: 21.0 MHZ - 21.450 MHZ.

	21.000 MHz	N/T CW	21.225 MHz	21.450 MHz
15	EX CW	CW RTTY DATA	EX VOICE	VOICE
	21.025 MHz	21.200 MHz		

The Extra Class operator has 2 exclusive operating segments on the 15 meter band. The bottom 25 kHz is reserved for the Extra Class operator. 21.200 to 21.225 MHz is also reserved for Extra Class operators, along with full privileges all the way up to the top of the band, 21.450 MHz.

The 15-meter band is loved by hams throughout the world because it has extremely low noise, and strong signal levels. There is little power line noise on 15 meters, and almost no atmospheric static. Band conditions on 15 meters usually favor daytime and evening contacts in the direction of the sun. Late at night, 15 meters begins to fade away, and you won't get skywave coverage until the next morning. 15 meters is a popular band for mobile operators because antennas are smaller.

12 METERS: 24.890 MHZ - 24.990 MHZ.

	24.890 MHz	24.930 MHz	24.990 MHz
12	CODE	VOICE	

The 12-meter band is a daytime band for the best long-range DX, and the Extra Class operator shares this entire band with General Class and grandfathered Advanced Class operators. CW starts at the bottom of the band at 24.890 MHz, and voice begins in the middle of the band at 24.930 MHz. There is plenty of excitement on the 12-meter band, and antenna requirements are smaller.

10 METERS: 28.0 MHZ - 29.7 MHZ.

28.000 MHz	28.500 MHz		29.700 MHz
N/T	N/T VOICE		
10 CW RTTY DATA	VOICE	VOICE	FM
28.000 MHz 28.300 MHz			29.500 MHz

There is plenty of excitement shared by all classes of amateur operators on the 10-meter band. Worldwide and continental band openings may pop up anytime there are extra amounts of solar activity. During the summertime, Sporadic-E ionospheric conditions make 1,500-mile band openings exciting for all operators on "10." The Extra Class operator has the entire 10-meter band from top to bottom, with plenty of CW activity at 28.1 MHz, propagation beacons at 28.2 MHz, phone excitement between 28.3 MHz to 29.0 MHz, and all sorts of elbow room, including FM operation, way up there between 29.5 and 29.7 MHz.

VOICE BAND EXPANSION

The Federal Communications Commission's WT Docket 04-140 Report and Order, issued in late 2006, amends Part 97 Rules to provide expanded voice privileges on the worldwide bands for General, Advanced and Extra Class operators. Novice and Technician licensees also were given additional CW privileges on the 80-, 40-, 15-, and 10-meter bands. This Report and Order is the FCC's response to "grossly underused... waste in spectrum..." within the CW portion of these bands. The Report and Order also removed the 200 watt power limitation for General, Advanced, and Extra Class operators in the CW portion of these bands.

On 75 meters, Generals may now operate voice from 3800 to 4000 kHz, an increase of 50 kHz. Advanced Class licensees are able to operate voice from 3700 to 4000 kHz, a 75 kHz increase. Extra Class operators gain 150 kHz of voice elbow room from 3600 to 4000 kHz.

On 40 meters, Advanced and Extra Class operators are able to use voice from 7125 to 7300 kHz, an increase of 25 kHz. General Class may now use from 7175 to 7300 kHz, a 50 kHz increase.

On 15 meters, Generals gain an additional 25 kHz of voice from 21.275 to 21.450 MHz.

Novice and Technician operators gain Morse code only privileges in an expanded portion of the lower half of 10-, 15-, 40-, and 80-meters.

Technician Class operators, who have never passed a code test, are now permitted 10 meter SSB voice communications from 28.300 MHz to 28.500 MHz, giving brand new ham operators a taste of skywave SSB voice communications on a portion of one worldwide band. As a new Extra, spend some time on "10" and encourage these new hams to upgrade!

New 60m Replacement Channel

One 60 meter channel, previously dial frequency 5366.5 kHz, is now replaced by dial frequency 5357.0 kHz, with an increase of power to 100 watts ERP. Multi-mode operation is now allowed on all 5 channels, too.

6 Meters and Up. Your Extra Class license allows you unlimited band privileges and unlimited emission privileges shared with all classes of license from Technician Class on up on the following bands:

Meters	Frequency
6 meters	50-54 MHz
2 meters	144-148 MHz
1.25 meters	222-225 MHz (219-220 MHz Digital)
70 centimeters	420-450 MHz
35 centimeters	902-928 MHz
23 centimeters	1240-1300 MHz

Microwave Bands. Many Extra Class operators are employed by companies that specialize in microwave electronics. The Extra Class license allows for unlimited privileges and unlimited emission privileges on the following microwave bands:

Frequency	Frequency
2300-2310 MHz	47.0-47.2 GHz
2390-2450 MHz	75.5-81.0 GHz
3.3-3.5 GHz	119.98-120.02 GHz
5.65-5.925 GHz	142-149 GHz
10.0-10.5 GHz (popular X band)	241-250 GHz
24.0-24.25 GHz	All above 300 GHz

Look for plenty of Extra Class excitement on the 10-GHz band. This may soon become the new frontier for those amateur operators establishing records on microwave frequencies.

BECOME A VOLUNTEER EXAMINER

Here's some additional information to further encourage you to become a Volunteer Examiner once you upgrade to Extra Class.

The Telecommunications Act of 1996 removed many of the burdensome conflict-of-interest, record keeping, and financial-certification requirements previously imposed on VEs and VECs (Volunteer Examiner Coordinators). As a result of these rule changes, there no longer are any regulatory prohibitions preventing VEs or VECs from distributing license preparation (study) materials (with or without profit) to anyone, or from being employed by a company that manufactures or sells amateur radio station equipment or license study materials. Amateurs who serve as ham class instructors may now distribute license preparation materials to their students and also serve as Volunteer Examiners for the course examinations. Serving as a VE is a great way to help our hobby grow!

BE AN EXTRA CLASS ELMER – TEACH A CLASS

When you pass your upcoming Extra Class exam, you will have arrived at the TOP! You can share your enthusiasm for our amateur radio service by becoming a Ham Instructor. The W5YI Ham Instructor program is designed to help spread the word – and training – to hobby radio enthusiasts to encourage them to earn their entry-level ham radio license.

Ham Instructors may find kids and scouting as a great place to introduce ham radio and develop amateur radio training classes. Boaters, flyers, and RVers are *all* interested in what amateur radio can do to make their traveling safer and more enjoyable. You might become an Elmer and help them locate club or local weekend classes to earn their entry-level Technician Class license.

The new Technician Class license is the ideal starting point for everyone wanting to learn about ham radio. No Morse code test involved! The Technician exam trains would-be hams on techniques of going on the air for the first time, and learning about all

**Become a Ham Ambassador
and help our hobby grow!**

the excitement ham radio has to offer, including repeaters and satellites.

As a W5YI Ham Instructor, the industry will support you with free band plan charts, frequency guides, and other training materials. Extra Class Ham Instructors help our service continually grow. To learn more about the W5YI Ham Instructor program, call 800-669-9594, and join the outreach program to bring more kids and radio enthusiasts into our hobby and public service. Visit **www.haminstructor.com**.

SUMMARY

Whether you are upgrading to Extra Class from the grandfathered Advance Class license or the General Class license, you will gain some exciting new privileges on shared and Extra-only voice and CW portions of the bands for worldwide communications. This is where all the rare DX happens, and you will be operating in an area of the lower portion of the sub-bands where General Class operators can only listen in with envy!

But your biggest privilege as an Extra Class operator is the ability for you to work closely with a Volunteer Examination Coordinator and serve as a VE team leader. Now that General Class operators, along with grandfathered Advanced Class operators, may also assist Extra Class operators in administering license exams, your roll as a contact Volunteer Examiner will be an important one for the growth of the testing teams – and the growth of our hobby – throughout the country.

2

A Little Ham History!

Ham radio has changed a lot in the 100+ years since radio's inception. In the past 15 years, we've seen some monumental changes! So, before we go any further, I'd like to give you a brief history of ham radio licensing in the U.S. After all, as an Extra Class operator administering exams, you should be able to answer those "rookie" questions about the history of our hobby.

In this chapter you'll learn all of the licensing requirements under the FCC rules that became effective April 15, 2000. And you'll learn about the six classes of license that were in effect *prior* to those rules changes. That way, when you run into a Novice, Technician Plus or Advanced class operator on the air, you'll have some understanding of their skill level, experience, and frequency privileges.

WHAT IS THE AMATEUR SERVICE?

There are more than 700,000 licensed amateur radio operators in the U.S. today. The Federal Communications Commission, the Federal agency responsible for licensing amateur operators, defines our radio service this way:

"The amateur service is for qualified persons of all ages who are interested in radio technique solely with a personal aim and without pecuniary interest."

Ham radio is first and foremost a fun hobby! In addition, it is a service. And note the word "qualified" in the FCC's definition – that's the reason why you're studying for an exam; so you can pass the exam, prove you are qualified, and get on the Extra Class bands.

Millions of operators around the world exchange ham radio greetings and messages by voice, teleprinting, telegraphy, facsimile, and television worldwide. Japan, alone, has more than a million hams! It is very commonplace for U.S. amateurs to communicate with Russian amateurs, while China is just getting started with its amateur service. Being a ham operator is a great way to promote international good will.

The benefits of ham radio are countless! Ham operators are probably known best for their contributions during times of disaster. In recent years, many recreational sailors in the Caribbean who have been attacked by modern-day pirates have had their lives saved by hams directing rescue efforts. Following the 9/11 terrorist attacks on the World Trade Center and the Pentagon, literally thousands of local hams assisted with emergency communications. In addition, over the years, amateurs have contributed much to electronic technology. They have even designed and built their own orbiting communications satellites.

If you study hard and make the effort, you are going to join the top echelon of our service – Extra Class! Follow the suggestions in my book and your chances of passing the written exam are excellent.

A BRIEF HISTORY OF AMATEUR RADIO LICENSING

Government licensing of radio stations and amateur operators began with The Radio Act of 1912, which mandated the first Federal licensing of all radio stations and assigned amateurs to the short wavelengths of less than 200 meters. These "new" requirements didn't deter them, and within a few years there were thousands of licensed ham operators in the United States.

Since electromagnetic signals do not respect national boundaries, radio is international in scope. National governments enact and enforce radio laws within a framework of international agreements, which are overseen by the International Telecommunications Union. The ITU is a worldwide United Nations agency headquartered in Geneva, Switzerland. The ITU divides the radio spectrum into a number of frequency bands, with each band reserved for a particular use. Amateur radio is fortunate to have many bands allocated to it all across the radio spectrum.

In the U.S., the Federal Communications Commission is the government agency responsible for the regulation of wire and radio communications. The FCC further allocates frequency bands to the various services in accordance with the ITU plan – including the Amateur Service – and regulates stations and operators.

In the early years of amateur radio licensing in the U.S., the classes of licenses were designated by the letters "A," "B," and "C." The highest license class with the most privileges was "A." In 1951, the FCC dropped the letter designations and gave the license classes names. They also added a new Novice class. In 1967, the Advanced class was added to the Novice, Technician, General and Extra classes. The General exam required 13-wpm code speed, and Extra required 20-wpm. Each of the five written exams were progressively more comprehensive and formed what came to be known as the *Incentive Licensing System.*

In 1979, the international Amateur Service regulations were changed to permit all countries to waive the manual Morse code proficiency requirement for "...stations making use exclusively of frequencies above 30 MHz." This set the stage for the creation of the Technician "no-code" license, which occurred in 1991, when the 5-wpm Morse code requirement for the Technician class was the eliminated.

By this time, there were a total of six Amateur Service license classes – Novice, Technician, Technician-Plus, General, Advanced, and Extra – along with five written exams and three Morse code tests used to qualify hams for their various licenses.

The Amateur Service Is Restructured

Following an extensive review begun in 1998, the FCC implemented a complete restructuring of the U.S. amateur service that became effective April 15, 2000. Today, applicants can only be examined for three amateur license classes:

- Technician class – the VHF/UHF entry level license;

- General class – the HF entry level license, and

- Amateur Extra class – a technically-oriented senior license.

Most recently, the FCC updated its amateur radio rules by eliminating the Morse code test requirement. The new rule went into effect in February, 2007, and it means that you can earn any class of license – Technician, General, and Extra class – simply by passing written exams.

Individuals with licenses issued before April 15, 2000, have been "grandfathered" under the new rules. This means that Novice and Advanced class amateurs are able to modify and renew their licenses indefinitely. Technician-Plus amateur licenses will be renewed as Technician class, and these licensees will retain their HF operating privileges indefinitely. The FCC elected not to change the operating privileges of any class, so you may hear some of these "grandfathered" hams when you get on the air.

Self-Testing In The Amateur Service

Prior to 1984, all amateur radio exams were administered by FCC personnel at FCC Field Offices around the country. In 1984, the VEC (Volunteer Examiner Coordinator) System was formed after Congress passed laws that allowed the FCC to accept the services of Volunteer Examiners (or VEs) to prepare and administer amateur service license examinations. The testing activity of VEs is managed by Volunteer Examiner Coordinators (or VECs). A VEC acts as the administrative liaison between the VEs who administer the various ham examinations and the FCC, which grants the license.

A team of three VEs, who must be approved by a VEC, is required to conduct amateur radio examinations.

In 1986, the FCC turned over responsibility for maintenance of the exam questions to the National Conference of VECs, which appointed a Question Pool Committee (QPC) to develop and revise the various question pools according to a schedule.

That completes your history lesson. If you'd like to learn more, visit The Ham Radio History forum at www.YahooGroups.com/list/Ham-Radio-History.

OPERATOR LICENSE CLASSES AND EXAM REQUIREMENTS

To qualify for an amateur operator/primary station license, a person must pass an examination according to FCC guidelines. The degree of skill and knowledge that the candidate demonstrates to the examiners determines the class of operator license for which the person is qualified.

Anyone is eligible to become a U.S. licensed amateur operator (including foreign nationals, if they are not a representative of a foreign government). There is no age limitation – if you can pass the examinations, you can become a ham!

Today, there are three amateur operator licenses issued by the FCC – Technician, General, and Extra. Each license requires progressively higher levels of learning and proficiency, and each gives you additional operating privileges. This is known as *incentive licensing* – a method of strengthening the amateur service by offering more radio spectrum privileges in exchange for more operating and electronic knowledge.

There is no waiting time required to upgrade from one amateur license class to another, nor any required waiting time to retake a failed exam. You can even take all three examinations at one sitting if you're really brave! *Table 3-1* details the amateur service license structure and required examinations.

Table 3-1: Current Amateur License Classes and Exam Requirements
(Effective April 15, 2000)

License Class	Exam Element	Type of Examination
Technician Class	2	35-question, multiple-choice written examination. Minimum passing score is 26 questions answered correctly (74%).
General Class	3	35-question, multiple-choice written examination. Minimum passing score is 26 questions answered correctly (74%).
Extra Class	4	50-question, multiple-choice written examination. Minimum passing score is 37 questions answered correctly (74%).

ABOUT THE WRITTEN EXAMS

What is the focus of each of the written examinations, and how does it relate to gaining expanding amateur radio privileges as you move up the ladder toward your Extra Class license? *Table 3-2* summarizes the subjects covered in each written examination element.

Table 3-2. Question Element Subjects

Exam Element	License Class	Subjects
Element 2	Technician	Elementary operating procedures, radio regulations, and a smattering of beginning electronics. Emphasis is on VHF and UHF operating.
Element 3	General	HF (high-frequency) operating privileges, amateur practices, radio regulations, and a little more electronics. Emphasis is on HF bands.
Element 4	Extra	Basically a technical examination. Covers specialized operating procedures, more radio regulations, formulas and heavy math. Also covers the specifics on amateur testing procedures.

No Jumping Allowed

All written examinations for an amateur radio license are additive. You *cannot* skip over a license class or by-pass a required examination as you upgrade from Technician to General to Extra. For example, to obtain a General Class license, you must first take and pass the Element 2 written examination for the Technician Class license, plus the Element 3 written examination. To obtain the Extra Class license, you must first pass the Element 2 (Technician) and Element 3 (General) written examinations, and then successfully pass the Element 4 (Extra) written examination.

TAKING THE ELEMENT 4 EXAM

Here's a summary of what you can expect when you go to the session to take the Element 4 written examination for your Extra Class license. Detailed information about how to find an exam session, what to expect at the session, what to bring to the session, and more, is included in Chapter 4.

All amateur radio examinations are administered by a team of at least three Volunteer Examiners (VEs) who have been accredited by a Volunteer Examiner Coordinator (VEC). The VEs are licensed hams who volunteer their time to help our hobby grow.

Examination sessions are organized under the auspices of an approved VEC. A list of VECs is located in the Appendix on page 269. The W5YI-VEC and the ARRL-VEC are the 2 largest examination groups in the country, and they test in all 50 states. Their 3-member, accredited examination teams are just about everywhere. So when you call the VEC, you can be assured they probably have an examination team only a few miles from where you are reading this book right now!

The Element 4 written exam is a multiple-choice format. The VEs will give you a test paper that contains the 50 questions and multiple choice answers, and an answer sheet for you to complete. Take your time! Make sure you read each question carefully and select the correct answer. Once you're finished, double check your work before handing in your test papers.

The VEs will score your test immediately, and you'll know before you leave the exam site whether you've passed. Chances are very good that, if you've studied hard, you'll get that passing grade!

Want to find a test site fast?
Visit the W5YI-VEC website at **www.w5yi.org**, or call 800-669-9594.

GETTING/KEEPING YOUR CALL SIGN

Once the VE team scores your exam and you've passed, the process of getting your upgrade FCC Amateur Radio Licenses begins – usually the same day.

At the exam site, you will complete NCVEC Form 605, which is your application to the FCC for your license upgrade. If you pass the exam, the VE team will send the required paperwork on to their VEC. The VEC reviews the paperwork submitted by your exam team and then files your application with the FCC. This filing is done electronically, and your upgrade license will be granted and posted on the FCC's website within a few days. As soon as you pass your exam, you are permitted to go on the air as a licensed Extra Class operator.

Should you keep your old call sign, or ask for a new one? If you check the box on the Form 605 asking for a new call sign, the FCC's computer will assign one to you. But you might not like what you get. So I suggest you stick with your old call sign, and if you want a new, specific, call file an application for a Vanity Call Sign. See Chapter 4 for more details.

IT'S EASY!

Probably the primary pre-requisite for passing any amateur radio operator license exam is the will to do it. If you follow my suggestions in this book, your chances of passing the Extra Class exam are excellent.

Yes, indeed, the year 2000 brought some big changes to ham radio, and everyone comes out a winner! There has never been a better time to join the top ranks of ham radio hobbyists. So study hard! We hope to hear you on the Extra Class bands very soon.

3

Getting Ready for the Exam

This chapter includes the entire, new Element 4 Question Pool questions and answers that can be used to make-up your 50-question Extra Class Element 4 written theory examination *exactly as they will appear on your exam.* Your exam will have 50 questions, and a score of 37 or more correctly-answered questions (74%) will earn you a passing grade – which means you cannot have more than 13 wrong. Answer at least 37 questions correctly, and you pass the Extra Class examination!

> THIS EXTRA CLASS QUESTION POOL IS VALID FROM
> JULY 1, 2012, THROUGH JUNE 30, 2016.

The Element 4 question pool contains a total of 702 active questions and multiple-choice answers and distracters (the false answers are called *distracters.*) 50 of these questions will appear on your upcoming Element 4 written examination. All of these examination questions, plus the precise multiple-choice answers – one of which is the correct answer – *are identical to those included in this book.*

The Volunteer Examiners are not permitted to reword the questions, nor are they permitted to change any of the right or wrong answers (but they can change the A B C D *order* of the answers). *Every question in this book, letter for letter, word for word, number for number, and every right and wrong answer, will be exactly the same on your upcoming 50-question Extra Class Element 4 exam.*

In this new 6th edition of my book, my emphasis is to educate you about Extra Class operating techniques, rules and regs, propagation, electrical theory and practical circuits, antennas, and more. I do this by placing the correct answer key at the very end of my explanation of the correct answer that appears after each question and 4 possible answers. In addition to some memory tricks to help you remember the correct answer, I also provide hints on how to identify wrong answers that often look very much like the correct one. Of course, my explanations will not appear on your test papers when you take your Element 4 written exam.

The new Element 4 Extra Class question pool was developed by the Question Pool Committee of the National Conference of Volunteer Exam Coordinators. They began by carefully reviewing the most recent Extra Class question pool, editing out more than 60 obsolete or duplicate questions, and then adding in about 60 brand new questions designed to keep the pool up-to-date with newer technology and current FCC rules and regulations. This new 2012-2016 question pool was carefully developed by the QPC under the leadership of chairman Roland Anders, K3RA, of the Laurel Amateur Radio Club VEC, and committee members Perry Green, WY1O, of the ARRL VEC, Michael Maston, N6OPH, of the San Diego Amateur Radio Club VEC, Larry Pollock, NB5X, of the W5YI VEC, and Jim Wiley, KL7CC, of the Anchorage Amateur Radio Club VEC.

Three dozen additional volunteers (including me!) also contributed potential test questions to the Question Pool Committee. We all want to thank the members of the Question Pool Committee and their Editor for their dedication and the hundreds of hours spent updating a very old Extra Class pool that contained some Q&As dating back to the '60s. Thanks, Team, for a fresh, new Extra Class question pool!

WHAT THE EXAM CONTAINS

Again, your exam will be comprised of 50 questions taken from the Element 4 question pool. The new pool is divided into 10 sub-elements. *Table 2-1* shows you the subelement topics, the total number of questions for each sub-element, and the number of questions from each subelement that will appear on your exam. Carefully study *Table 2-1* – it will help you understand how the questions are distributed among the 10 sub-elements and how your exam will be constructed.

Table 2-1. Question Distribution for the Extra Class Element 4 Exam

Subelement Number	Subelement Topic	No. of Questions in Pool	No. of Questions on Exam
E1	Commission's Rules (FCC rules for the Amateur Radio services)	74	6
E2	Operating Procedures (Amateur station operating procedures)	68	5
E3	Radio Wave Propagation (Radio wave propagation characteristics)	35	3
E4	Amateur Practices	70	5
E5	Electrical Principles (Electrical principles as applied to amateur station equipment)	71	4
E6	Circuit Components (Amateur station equipment circuit components)	83	6
E7	Practical Circuits (Practical circuits employed in amateur station equipment)	125	8
E8	Signals and Emissions (Signals and emissions transmitted by amateur stations)	56	4
E9	Antennas and Feed Lines (Amateur station antennas and feed lines)	109	8
E0	Safety	11	1
Total		702	50

(Titles in parentheses are the subelement titles on which you will be examined.)

NOTE: *The Element 4 Question Pool originally contained 703 questions. One question was deleted by the QPC prior to publication of this book, resulting in a pool with 702 active questions. The deleted question does not appear in this book*

QUESTION POOL VALID THROUGH JUNE 30, 2016

The Element 4 question pool in this book is valid from July 1, 2012, until June 30, 2016. During this period, the questions and answers are frozen. The exact numerical values found in each Extra Class question and answer will remain precisely the same. If FCC rules change or advancements in technology cause any question to become obsolete, that question will be eliminated from your test and there will be no surprises in the exam room.

The new 2012-16 Extra Class Element 4 pool covers many new subject areas designed to keep the Extra Class questions fresh:

- *Ham Rules and Regulations on band edge SSB operation*
- *Satellite operation and modes*
- *ATV and SSTV technicalities*
- *Contest operating*
- *Contest log files*
- *PSK and QPSK*
- *Antenna takeoff patterns*
- *Antenna analyzer versus SWR bridge*
- *NPN transistors*
- *Impedance calculations*
- *Q for a series tuned circuit*
- *dBm noise floor calculations*
- *Roofing filter topics*

- *Receiver blocking dynamic range*
- *Intermodulation interference*
- *Third order intercept levels*
- *Noise blankers*
- *RF attenuators*
- *Schottky barrier diodes*
- *Emitter amplifier bias*
- *Third order intermodulation distortion*
- *Parametric amplifiers*
- *High voltage power supply filters*
- *Frequency counters*
- *Antenna modeling*
- *Safety practices, RF radiation hazards, hazardous materials*

QUESTION CODING

Each question in the Element 4 Extra Class pool is assigned a coded number using a system of numbers and letters. *Figure 2-1* explains the coding for question E5G09. "E" stands for Extra Class Element 4 question pool. "5" indicates subelement 5, Electrical Principles. "G" indicates the sub-element group on Circuit Q; Reactive Power; Power Factor. "09" indicates the ninth question dealing with this topic.

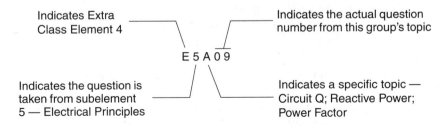

Indicates Extra Class Element 4

Indicates the actual question number from this group's topic

E 5 A 0 9

Indicates the question is taken from subelement 5 — Electrical Principles

Indicates a specific topic — Circuit Q; Reactive Power; Power Factor

Figure 2-1. Examination Question Coding

If you carefully study the numbering scheme, you will find that each subelement is divided into topic groups, and there will be one question on your examination taken from each topic group. If there is a specific group of questions which are giving you a hard time, study them later. Only one question will be taken from any one group.

To facilitate in identifying subelement groups, we describe the subelement contents by individual groups of questions. This is good news for those of you who may be a bit rusty on a specific topic, knowing ahead of time that your upcoming Extra Class examination will only have one question out of each topic group.

STUDY HINTS

You're probably saying: "702 questions and answers is an overwhelming number of questions and answers to memorize for the new Extra Class exam!" But good news – almost every question is asked at least twice with just slightly different phrasing. Many of the math questions are based on single-formulas; and once you know the formula, you can easily answer the 4 or 5 questions based on that formula.

The mathematical formulas that you will need for the upcoming Extra Class exam are noted with an ▶ in the question pool that follows, and they are summarized in the Appendix on pages 274 to 278. Memorize the formulas, understand how to use them, and make sure you know the correct units to use. Many times you may have to convert kilohertz (kHz) to megahertz (MHz), or vice versa.

Work your book! Put a check mark beside questions you have down cold, and circle those that may need a little bit more study. Highlight those that appear to be brain-busters, and figure out ways to memorize the correct answers. Begin to memorize the formulas, and learn how to apply them to arrive at a correct answer. In many cases, there are handy shortcuts to help you through the longer formulas, and a little humor now and then should keep you on track.

If you study the questions and answers for a half hour in the morning, and a half hour in the evening, it should take you approximately 30 days to prepare for the Extra Class exam. If you are a grandfathered Advanced Class operator, it may take you only 15 days to prepare for your upgrade. One week before the exam, highlight each question's correct answer and spend that week only looking at the question and highlighted correct answer.

QUESTIONS REARRANGED FOR SMARTER LEARNING

The first thing you'll notice when you look at how the Element 4 question pool is presented in my *Extra Class* book is that I have completely rearranged the entire question pool into logical topic groups for easier learning. This rearrangement is the same presentation I use when I teach my 4-day Extra Class seminars. Rather than having to jump back and forth between subelement topic areas to study similar-topic questions, I take you LOGICALLY through the entire question pool with all of the Q&A arranged into 16 topic areas. For example, in the numerically-ordered question pool as issued by the QPC, antenna questions are sprinkled throughout various subelements. The same thing is true of questions about transmit bandwidth and many other topics. Here is a list of the topic areas and the page numbers on which each begins in this book:

We provide a numerical cross-reference of all of the Element 4 questions in the Appendix, on page 270, so if you need to look up a specific question by its number, we tell you precisely the page that it is on. This logical re-arrangement of questions is an absolute exclusive training method with all Gordon West ham radio test preparation books.

Just as in my *Technician Class* and *General Class* books, I have taken every single question in the Element 4 pool and placed it within the proper topic area for logical learning as a group. My classroom students love this approach to learning, and you will, too! Key words are also highlighted in *blue* in the explanations.

ADDITIONAL STUDY MATERIALS

Passing your Extra Class exam will require plenty of Q&A study. You can back-up the material in my book with some additional resources that will help you learn – and understand – the material.

First, there is my 6 CD audio course for Extra Class that is an excellent addition to this book. If you spend a lot of time in your car or pickup, or if you just learn better by listening, my audio CD course is for you. It follows the same outline as this book, and I include many sounds of Extra Class operation to help bring the material to life.

Next, if you like learning at your PC, study software is available that accompanies my book. This W5YI program includes the answer explanations from my book. When you take a practice exam, the software shows you the correct answer, scores your work, and tells you where you need to focus your study efforts.

You can back up all of this material presented in my book, audio course, and software program the well-illustrated "basic" books we offer on electronics – *Basic Electronics, Basic Communications Electronics,* and *Basic Digital Electronics* – all of which are available where you purchased this book, or from The W5YI Group. If you are extra technical, we also have 4 Engineer's Mini Notebooks written by Forrest M. Mims III, which parallel great electronic projects with step by step instructions. Check out these additional study materials in the back of this book.

Of course, the real learning of the material covered in these questions happens after you pass your exam and get on the air as an Amateur Extra Class operator!

EXAMINATION QUESTIONS

Let's get started reviewing the 702 Q&As, 50 of which will be chosen for your upcoming Extra Class exam. Read over the question and see if you already know the answer. Next, spot the likely answer and then read my explanation of the correct answer. Finally, at the end of the explanation, I give you the correct A, B, C, or D answer. These Q&As are exactly the same as those found on your upcoming Extra Class exam – verbatim. Use a small card to cover up the explanations and then slowly reveal the 4 possible answers, the explanation, and then the correct answer. Put a check mark beside the question number if you answered it correctly. Put an X if you need more study with a specific question.

Work the book! In our classes, I can usually tell who will pass the exam the first time by looking at their marked up textbooks.

Ready to get started? Here we go with the first series of questions.

Rules & Regs

E1A01 When using a transceiver that displays the carrier frequency of phone signals, which of the following displayed frequencies represents the highest frequency at which a properly adjusted USB emission will be totally within the band?

    A. The exact upper band edge.
    B. 300 Hz below the upper band edge.
    C. 1 kHz below the upper band edge.
    D. 3 kHz below the upper band edge.

Worldwide high frequency transceivers may automatically select the correct upper sideband or lower sideband when you start jumping around the bands. (Newer sideband transceivers also include the upper sideband requirement for the 60 meter band.) Let's say you are on 20 meters and dialed in to the maritime mobile net frequency of 14.300 MHz. This is the CARRIER frequency, and on 20 meters our 14.300 MHz voice emission will modulate UP approximately 2.8 kHz. Your transmitted voice will then occupy just under 3 kHz of bandwidth, spanning from 14.300 to 14.303 MHz. The test question and correct answer illustrate the requirement that any upper sideband signal must have the carrier frequency displayed on your dial at least *3 kHz below the upper band edge*. 14.347 MHz would be the closest we would want to dial our carrier frequency to stay within the upper limit of the 20 meter band. [97.301, 97.305] **ANSWER D.**

E1A03 With your transceiver displaying the carrier frequency of phone signals, you hear a DX station's CQ on 14.349 MHz USB. Is it legal to return the call using upper sideband on the same frequency?
- A. Yes, because the DX station initiated the contact.
- B. Yes, because the displayed frequency is within the 20 meter band.
- C. No, my sidebands will extend beyond the band edge.
- D. No, USA stations are not permitted to use phone emissions above 14.340 MHz.

No! Every so often you may hear a DX station calling CQ too close to a band edge. Don't be tempted to operate out of band by telling them that they are too close to the band edge! *Band edge operation may result in sidebands extending beyond the USA band edge limits*, which is a rules violation for us. [97.301, 97.305] **ANSWER C.**

E1A02 When using a transceiver that displays the carrier frequency of phone signals, which of the following displayed frequencies represents the lowest frequency at which a properly adjusted LSB emission will be totally within the band?
- A. The exact lower band edge.
- B. 300 Hz above the lower band edge.
- C. 1 kHz above the lower band edge.
- D. 3 kHz above the lower band edge.

This question now has us on lower sideband, maybe 40 meters, so we need to *stay 3 kHz ABOVE the lower band edge* to stay within the band. It would be considered poor operating practice to intentionally hug a band edge, and switch to the non-conforming sideband. [97.301, 97.305] **ANSWER D.**

E1A04 With your transceiver displaying the carrier frequency of phone signals, you hear a DX station calling CQ on 3.601 MHz LSB. Is it legal to return the call using lower sideband on the same frequency?
- A. Yes, because the DX station initiated the contact.
- B. Yes, because the displayed frequency is within the 75 meter phone band segment.
- C. No, my sidebands will extend beyond the edge of the phone band segment.
- D. No, USA stations are not permitted to use phone emissions below 3.610 MHz.

On 75 meters, we use lower sideband, and the lowest frequency near the voice band edge that we would select as our dial carrier frequency is 3.603 MHz. USA operators on *3.601 MHz* are extending *illegally beyond the USA band edge* of the phone segment. [97.301, 97.305] **ANSWER C.**

E1A12 With your transceiver displaying the carrier frequency of CW signals, you hear a DX station's CQ on 3.500 MHz. Is it legal to return the call using CW on the same frequency?
- A. Yes, the DX station initiated the contact.
- B. Yes, the displayed frequency is within the 80 meter CW band segment.
- C. No, sidebands from the CW signal will be out of the band.
- D. No, USA stations are not permitted to use CW emissions below 3.525 MHz.

Be cautious NOT to operate precisely at the band edge of your privileges. Emissions of your *CW* transmitted *signal might extend beyond band edge* limitations. [97.301, 97.305] **ANSWER C.**

E1A07 What is the only amateur band where transmission on specific channels rather than a range of frequencies is permitted?

A. 12 meter band.
B. 17 meter band.
C. 30 meter band.
D. 60 meter band.

It is only on the *60 meter band* where our ham radio transmissions are limited to 5 specific channels. [97.303] **ANSWER D.**

60 METERS	5 SPECIFIC CHANNELS – 100 WATTS ERP					VOICE SIDEBAND USB
	5330.5 kHz	5346.5 kHz	5357.0 kHz	5371.5 kHz	5403.5 kHz	

E1A06 Which of the following describes the rules for operation on the 60 meter band?

A. Working DX is not permitted.
B. Operation is restricted to specific emission types and specific channels.
C. Operation is restricted to LSB.
D. All of these choices are correct.

We share our 60 meter band with primary Federal stations. As secondary users of the *60 meter band*, we must not cause harmful interference to (and must accept interference from) stations authorized by the United States (NTIA and FCC) and other nations in the fixed service, and other nations in the mobile service (except aeronautical mobile service). We've been allocated *five channels* within the 60 meter band, and recent rule changes now allow 100 watts ERP (referenced to a half wave dipole), upper sideband using *voice, CW, and data* using PACTOR III techniques and RTTY using the PSK31 technique. [97.303] **ANSWER B.**

CARRIER TUNING FREQUENCY	CENTER FREQUENCY
5330.5 kHz	5332.0 kHz
5346.5kHz	5348.0 kHz
5357.0 kHz (New replacement)	5358.5 kHz
5371.5 kHz	5373.0 kHz
5403.5 kHz	5405.0 kHz

Recently, the previously authorized channel 5368 kHz was replaced with a new center frequency, 5358.5 kHz. Amateur radio operators also now have multiple modes authorized on the five 60-meter channels – voice, CW, and data modes using PACTOR III and RTTY using the PSK31 technique. Here's the hard part – we are required to share all of this on these five 60 meter channels. The ARRL recommends digital modes PSK31 and PACTOR III only. Time-sharing the channels, and cooperating with other stations that also have access to these channels, may allow these multiple modes to co-exist on the 60 meter band, which is perfectly situated in the spectrum for effective emergency communications. Learn more about the 60 meter regs on the FCC website: www.fcc.gov/document/amateur-radio-service-5-mhz.
And learn more by visting: www.arrl.org/news/fcc-releases-new-rules-for-60-meters.

E1A05 What is the maximum power output permitted on the 60 meter band?

A. 50 watts PEP effective radiated power relative to an isotropic radiator
B. 50 watts PEP effective radiated power relative to a dipole
C. 100 watts PEP effective radiated power relative to the gain of a half-wave dipole
D. 100 watts PEP effective radiated power relative to an isotropic radiator

The FCC recently granted higher power output on the 60 meter band. We can now transmit *100 watts* effectively radiated power *(ERP) relative to the gain of a half-wave dipole*. Since most HF rigs output just *100 watts*, you are good to go with the dipole 60 meter antenna. However, running an amplifier on 60 meters to exceed 100 watts output is not allowed. [97.313] **ANSWER C.**

E1B05 What is the maximum bandwidth for a data emission on 60 meters?

A. 60 Hz.
C. 1.5 kHz.
B. 170 Hz.
D. 2.8 kHz.

FCC Rule 97.303 (h) states that amateur radio operators must insure that *emissions from their 60 meter stations do not occupy more than 2.8 kHz bandwidth*, centered on the channel center frequency. PSK31 audio is on channel center frequencies, generally 1500 Hz above the suppressed carrier frequency. However, CW, 150HA1A must be centered on channel, where channel sharing and cooperation are of paramount importance. [97.303] **ANSWER D.**

E1B07 Where must the carrier frequency of a CW signal be set to comply with FCC rules for 60 meter operation?

A. At the lowest frequency of the channel.
B. At the center frequency of the channel.
C. At the highest frequency of the channel.
D. On any frequency where the signal's sidebands are within the channel.

The CW designator 150HA1A precludes tone modulated CW, MCW, where different tones could help sort out different CW stations transmitting on the same channel. The new rules read: "For *CW emissions, the carrier frequency is set to the CENTER frequency*." This means that if all ham sets sending CW have zero error on the center channel display, all the CW signals will come out with the same exact tone! Band planning may help solve this problem of multiple CW signals sharing a specific channel sounding the same! [97.15] **ANSWER B.**

E1A13 Who must be in physical control of the station apparatus of an amateur station aboard any vessel or craft that is documented or registered in the United States?

A. Only a person with an FCC Marine Radio.
B. Any person holding an FCC-issued amateur license or who is authorized for alien reciprocal operation.
C. Only a person named in an amateur station license grant.
D. Any person named in an amateur station license grant or a person holding an unrestricted Radiotelephone Operator Permit.

If you go offshore aboard your U.S. documented or registered boat, *anyone* on board *holding an FCC issued amateur radio license* of the correct grade can work the ham radio equipment. This also holds true for your non-U.S.-citizen friend who holds an *FCC-issued reciprocal license*. But just because you may be in International waters, non-licensed personnel (the ship's Captain, for instance) would not be permitted to operate on the ham bands, even though they may hold a Marine Radiotelephone Operator's Permit. [97.5(a)] **ANSWER B.**

E1A11 What authorization or licensing is required when operating an amateur station aboard a U.S.-registered vessel in international waters?
   A. Any amateur license with an FCC Marine or Aircraft endorsement.
   B. Any FCC-issued amateur license or a reciprocal permit for an alien amateur licensee.
   C. Only General class or higher amateur licenses.
   D. An unrestricted Radiotelephone Operator Permit.

This is a great question dealing with maritime mobile operation when out on the high seas. If the vessel is flying the American flag, registered in the United States, *any ham radio transmissions MUST be from an FCC-licensed amateur operator or person holding a reciprocal permit for an alien amateur license*. Years ago there was the false belief that once someone was in international waters, anyone could transmit over ham radio without a license. Not so – your ham license is required, and is separate from the ship's marine radio license or the captain's Marine Radio Operator Permit. [97.5] **ANSWER B.** ☞ Visit: www.arrl.org/regulations

E1A10 If an amateur station is installed aboard a ship or aircraft, what condition must be met before the station is operated?
   A. Its operation must be approved by the master of the ship or the pilot in command of the aircraft.
   B. The amateur station operator must agree to not transmit when the main ship or aircraft radios are in use.
   C. It must have a power supply that is completely independent of the main ship or aircraft power supply.
   D. Its operator must have an FCC Marine or Aircraft endorsement on his or her amateur license.

If you are planning a long cruise on your pal's boat, or taking a ride in his airplane, make sure you have received *approval by the captain (master of the ship) or aircraft pilot before you begin to transmit*. Look at the other 3 incorrect answers – they sound logical, but only the answer we just gave you is correct. [97.11] **ANSWER A.**

E1B03 Within what distance must an amateur station protect an FCC monitoring facility from harmful interference?
   A. 1 mile.   C. 10 miles.
   B. 3 miles.   .   D. 30 miles.

The Federal Communications Commission recently automated its monitoring stations. Now there are remote-controlled, unmanned monitoring stations, and all amateur *operators within one mile of those stations must contact the local FCC engineer in charge* to ensure they do not cause harmful interference to the remote-controlled monitoring units. [97.13] **ANSWER A.**

E1F06 What is the National Radio Quiet Zone?
   A. An area in Puerto Rico surrounding the Aricebo Radio Telescope.
   B. An area in New Mexico surrounding the White Sands Test Area.
   C. An area surrounding the National Radio Astronomy Observatory.
   D. An area in Florida surrounding Cape Canaveral.

A national *radio quiet zone* has been established near the *National Radio Astronomy Observatory* covering parts of Maryland, West Virginia, and Virginia. The quiet zone restriction applies to "constant-on" stations like beacons and repeaters. Operation from your home or vehicle would not be affected unless you actually enter the Observatory property in Greenbank, WV. [97.3] **ANSWER C.**

E1B02 Which of the following factors might cause the physical location of an amateur station apparatus or antenna structure to be restricted?
  A. The location is near an area of political conflict.
  B. The location is of geographical or horticultural importance.
  C. The location is in an ITU zone designated for coordination with one or more foreign governments.
  D. The location is of environmental importance or significant in American history, architecture, or culture.

The *environmental impact* of a proposed installation is now a big consideration on anything that may have significant *impact on American history, architecture, or culture*. For instance, you would not want to place a station on an Indian burial ground. [97.13] **ANSWER D.** ☞ **Visit: www.fcc.gov/mb/facts**

E1B04 What must be done before placing an amateur station within an officially designated wilderness area or wildlife preserve, or an area listed in the National Register of Historical Places?
  A. A proposal must be submitted to the National Park Service.
  B. A letter of intent must be filed with the National Audubon Society.
  C. An Environmental Assessment must be submitted to the FCC.
  D. A form FSD-15 must be submitted to the Department of the Interior.

If you plan to install your station on a wildlife preserve, or an area listed as an historical place, you will need to *submit an environmental assessment to the FCC* for its review and approval. Be prepared for an extremely long wait before your permit may be granted! [97.13, 1.1305-1.1319] **ANSWER C.**

E1F13 What types of communications may be transmitted to amateur stations in foreign countries?
  A. Business-related messages for non-profit organizations.
  B. Messages intended for connection to users of the maritime satellite service.
  C. Communications incidental to the purpose of the amateur service and remarks of a personal nature.
  D. All of these choices are correct.

With your new Extra Class privileges you will likely make even more foreign country contacts. Remember, no business communications allowed, so keep your comms to purely *personal remarks, or communications of a technical nature*, when talking with hams in foreign countries. [97.117] **ANSWER C.**

Good To Know:

Going "down under?" The Australian Communications and Media Authority (ACMA) has issued a new class of license to allow visiting amateurs to operate in Australia for up to 90 days. All you have to do is use your home call sign followed by the suffix "VK" followed by "portable" and then announce the location of your station, without doing anything more.

E1F02 Which of the following operating arrangements allows an FCC-licensed U.S. citizen to operate in many European countries, and alien amateurs from many European countries to operate in the U.S.?

A. CEPT agreement.　　　　　C. ITU reciprocal license.
B. IARP agreement.　　　　　D. All of these choices are correct.

The *CEPT agreement* (European Conference of Postal & Telecommunications Administrations – CEPT) is great news for all U.S. Advanced and Extra Class amateur operators *when visiting many countries in Europe* – NO MORE RECIPROCAL PAPERWORK! Before leaving for Europe, obtain a copy of FCC Public Notice DA99-1098, released June 7, 1999, along with the original and copies of your U.S. ham license. This paperwork is necessary in case you are asked about the equipment you are bringing into that country at customs. Have fun in Europe with ham radio! Don't forget your paperwork and the required FCC Public Notice. [97.5] **ANSWER A.**

Participating CEPT countries, as of December, 2011, are:

Albania	Finland	Lithuania	Russian Federation
Austria	France & its	Luxembourg	Serbia
Belgium	possessions	Macedonia	Slovak Republic
Bosnia &	Germany	Monaco	Slovenia
Herzegovina	Greece	Montenegro	Spain
Bulgaria	Hungary	Netherlands	Sweden
Croatia	Iceland	Netherland Antilles	Switzerland
Cyprus	Ireland	Norway	Turkey
Czech Republic	Italy	Poland	Ukraine
Denmark	Latvia	Portugal	United Kingdom &
Estonia	Liechtenstein	Romania	its possessions

Good To Know: European Reciprocal Licenses Now Limited to Advanced & Extra Class Licensees

The European Conference of Postal and Telecommunications Administrations (CEPT) has revised its table of equivalence between FCC amateur licenses and the CEPT license. Effective February 4, 2008, Recommendation T/R 61-01 (as amended) now grants **full CEPT privileges only to those U.S. citizens who hold an FCC-issued Advanced or Amateur Extra Class license**. This means that those U.S. licensees who hold an FCC-issued General or Technician license are no longer eligible for full operating privileges in countries where CEPT-reciprocal operation had previously been permitted. U.S. Novice class licensees have had no reciprocal operating privileges under the CEPT provisions.

E1F04 Which of the following geographic descriptions approximately describes "Line A"?

A. A line roughly parallel to and south of the US-Canadian border.
B. A line roughly parallel to and west of the U.S. Atlantic coastline.
C. A line roughly parallel to and north of the US-Mexican border and Gulf coastline.
D. A line roughly parallel to and east of the U.S. Pacific coastline.

The *United States and Canada* have adopted the *"Line A" as a buffer zone* for certain types of radio stations. This keeps normally "line of sight" radio energy from interfering with the other country's communications. Amateurs are restricted on the 70-cm band between 420-430 MHz from transmitting if they are located north of "Line A." This is because Canada has a different type of radio allocation in the bottom part of the 70-cm band. [97.3] **ANSWER A.**

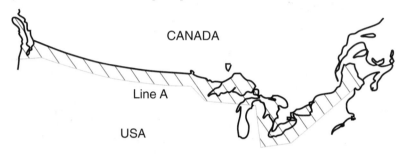

Amateur Radio Restrictions

E1F05 Amateur stations may not transmit in which of the following frequency segments if they are located in the contiguous 48 states and north of Line A?

A. 440 - 450 MHz.
B. 53 - 54 MHz.
C. 222 - 223 MHz.
D. 420 - 430 MHz.

Frequencies between 420-430 MHz are normally used for digital modes, fast-scan television, and auxiliary links. *Within the "Line A" area along the Canadian border, U.S. hams are prohibited from operating below 430 MHz.* [97.303] **ANSWER D.**

E1B06 Which of the following additional rules apply if you are installing an amateur station antenna at a site at or near a public use airport?

A. You may have to notify the Federal Aviation Administration and register it with the FCC as required by Part 17 of FCC rules.
B. No special rules apply if your antenna structure will be less than 300 feet in height.
C. You must file an Environmental Impact Statement with the EPA before construction begins.
D. You must obtain a construction permit from the airport zoning authority.

Live right next to an airport? If you do, you are going to need to *consult Part 17 of the FCC Rules* regarding the installation of that tall structure. Likely, you will be *working with the Federal Aviation Administration* to get that tower sanctioned, as well as documented, under Part 17 Rules. [97.15] **ANSWER A.**

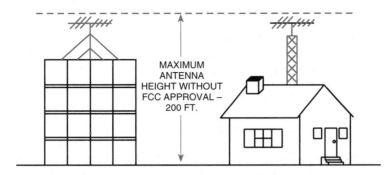

Antenna Height near a public use airport is governed by both the FCC and the FAA.
To learn more about the FAA rules and regs, visit its website.

E1F07 When may an amateur station send a message to a business?
A. When the total money involved does not exceed $25.
B. When the control operator is employed by the FCC or another government agency.
C. When transmitting international third-party communications.
D. When neither the amateur nor his or her employer has a pecuniary interest in the communications.

In 1993, the prohibited business communications rules were redefined by the FCC to allow an amateur to *send a message to a business as long as neither the amateur nor his or her employer has a pecuniary interest in the communications.* If you run a tree-trimming business, it would not be legal to use an autopatch to call your answering machine to find out where your next customer is waiting. On the other hand, if you are a school teacher and discover that a big pine tree has fallen down across the school's driveway, it would be perfectly acceptable to place an autopatch call to a tree-trimmer and have them come out to do their job. But could the school teacher call into the school district and tell them that she is going to be late for work? Probably NOT legal. [97.113] **ANSWER D.**

E1F08 Which of the following types of amateur station communications are prohibited?
A. Communications transmitted for hire or material compensation, except as otherwise provided in the rules.
B. Communications that have a political content, except as allowed by the Fairness Doctrine.
C. Communications that have a religious content.
D. Communications in a language other than English.

This question deals with permissible communications from one ham to another. Political, religious, and foreign language communications ARE permitted, but *communications transmitted for hire or material compensation are normally prohibited.* [97.113] **ANSWER A.**

E1A08 If a station in a message forwarding system inadvertently forwards a message that is in violation of FCC rules, who is primarily accountable for the rules violation?
A. The control operator of the packet bulletin board station.
B. The control operator of the originating station.
C. The control operators of all the stations in the system.
D. The control operators of all the stations in the system not authenticating the source from which they accept communications.

A few years ago, dozens of amateur operators were sent "greetings" from the FCC for taking part in forwarding a packet radio message that contained materials specifically prohibited by the rules and regulations. The hams forwarding the messages claimed everything was in the automatic mode and it was not up to them to censor what was coming down the electronic data line! Hence, new rules hold the *control operator of the station originating the message as primarily accountable* for any rule violation, and the control operator of the first forwarding station also must accept accountability for any violation of the rules. More specifically, everyone else "down the line" is no longer responsible for the message content, but must shut down their packet bulletin boards if informed that there is a message being automatically forwarded in violation of the rules. [97.219] **ANSWER B.**

E1A09 What is the first action you should take if your digital message forwarding station inadvertently forwards a communication that violates FCC rules?
A. Discontinue forwarding the communication as soon as you become aware of it.
B. Notify the originating station that the communication does not comply with FCC rules.
C. Notify the nearest FCC Field Engineer's office.
D. Discontinue forwarding all messages.

If you are advised that your packet station is automatically forwarding communications in violation of the rules, get into the computer and *discontinue forwarding this specific communication* as soon as you become aware of it. You may continue to forward all other messages. [97.219] **ANSWER A.**

E1C07 What is meant by local control?
A. Controlling a station through a local auxiliary link.
B. Automatically manipulating local station controls.
C. Direct manipulation of the transmitter by a control operator.
D. Controlling a repeater using a portable handheld transceiver.

Local control is the actual place where you turn on your equipment, press the microphone and transmit. This is called *"direct manipulation of the transmitter"* by a control operator. [97.3] **ANSWER C.**

E1C01 What is a remotely controlled station?
A. A station operated away from its regular home location.
B. A station controlled by someone other than the licensee.
C. A station operating under automatic control.
D. A station controlled indirectly through a control link.

Popular amateur radio IRLP stations located high atop mountains or buildings do not have the control operator sitting right in the repeater room. The station is controlled through a control link station. Most control link frequencies are found on

UHF 420-430 MHz, or up on 1.2 GHz. *"Control link" is the key for remote control* of an amateur station or repeater. [97.3] **ANSWER D.**

E1C08 What is the maximum permissible duration of a remotely controlled station's transmissions if its control link malfunctions?
 A. 30 seconds.
 B. 3 minutes.
 C. 5 minutes.
 D. 10 minutes.

If a distant, remotely-controlled station is undergoing remote control tone commands from an auxiliary station, and the control link on microwave fails, that remotely-controlled station must *shut down within three minutes*. [97.213] **ANSWER B.**

E1C06 Which of the following statements concerning remotely controlled amateur stations is true?
 A. Only Extra Class operators may be the control operator of a remote station.
 B. A control operator need not be present at the control point.
 C. A control operator must be present at the control point.
 D. Repeater and auxiliary stations may not be remotely controlled.

To *remotely control* another ham station, the *control operator must be present at the control point*. Only under AUTOMATIC control, like a repeater, may the control operator get some sleep! [97.109] **ANSWER C.**

E1C02 What is meant by automatic control of a station?
 A. The use of devices and procedures for control so that the control operator does not have to be present at a control point.
 B. A station operating with its output power controlled automatically.
 C. Remotely controlling a station's antenna pattern through a directional control link.
 D. The use of a control link between a control point and a locally controlled station.

Ham operators have come up with *devices and procedures* so they don't have to stand guard at the repeater block house on a mountain top. [97.3, 97.109] **ANSWER A.**

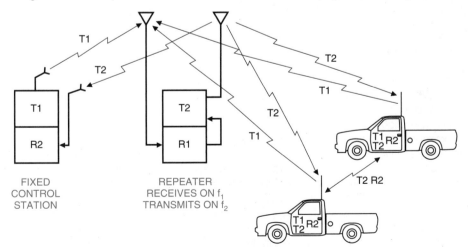

Repeater
Source: *Mobile 2-Way Radio Communications*, G. West, Copyright ©1992 Master Publishing, Inc., Niles, IL

E1C04 When may an automatically controlled station retransmit third party communications?

A. Never.
B. Only when transmitting RTTY or data emissions.
C. When specifically agreed upon by the sending and receiving stations.
D. When approved by the National Telecommunication and Information Administration.

A good example of third party messages from an automatically-controlled station is PACKET radio. *PACKET, along with RTTY, is a data emission* which has specific spots reserved on the high frequency radio dial when *operated in the automatic mode*. For instance, on 15 meters, an automatically controlled digital station has 10 kHz at 21.090 – 21.100 MHz. On 20 meters, the automatically controlled station may only pass third party communications between 14.1005 -14.1120 MHz. These automatic digital stations are unattended, but there still must be a control operator who will ensure absolute compliance with the rules and regulations. This automatically controlled RTTY or data emission will respond to interrogation by a ham station that is under local or remote control. Also, the automatically-controlled station may not occupy a bandwidth of more than 500 Hz. The very popular Winlink 2000 running relatively narrow PACTOR 1 and PACTOR 2 may be found anywhere below the SSB bands, but the wider PACTOR 3 has a control operator to initiate an unattended "autostart" station under FCC rule [97.109]. **ANSWER B.**

E1C03 How do the control operator responsibilities of a station under automatic control differ from one under local control?

A. Under local control there is no control operator.
B. Under automatic control the control operator is not required to be present at the control point.
C. Under automatic control there is no control operator.
D. Under local control a control operator is not required to be present at a control point.

Repeaters under *automatic control don't have anyone sitting up there* at that distant control point. [97.3, 97.109] **ANSWER B.** ☞ **Visit: www.winlink.org**

E1C05 When may an automatically controlled station originate third party communications?

A. Never.
B. Only when transmitting an RTTY or data emissions.
C. When specifically agreed upon by the sending and receiving stations.
D. When approved by the National Telecommunication and Information Administration.

An example of third party and third party traffic would be messages from one control operator to another ham station control operator on behalf of a close friend or relative who may not have their own ham license. If we read this question correctly, it asks "may an automatically controlled station ORIGINATE third party communications." The answer is *NEVER*, because a third party without a ham license may not be designated by the first control operator to assume the control operator function required to ORIGINATE that third party communication. Even though the station may be automatically controlled and sending data emissions, a third party is not allowed to electronically key up your station and begin messaging. It takes a licensed control operator to initiate an unattended "auto-start" station that may be handling third party traffic. [97.109] **ANSWER A.**

E1C10 What types of amateur stations may automatically retransmit the radio signals of other amateur stations?
A. Only beacon, repeater or space stations.
B. Only auxiliary, repeater or space stations.
C. Only earth stations, repeater stations or model craft.
D. Only auxiliary, beacon or space stations.

Automatically retransmitting the radio signals of other ham stations takes place with a common *repeater*, and can also take place from *satellite and space stations*, and can also take place from an *auxiliary station*, such as one handling IRLP. [97.113] **ANSWER B.** ☞ **Visit: www.winsystem.org**

E1C09 Which of these frequencies are available for an automatically controlled repeater operating below 30 MHz?
A. 18.110 - 18.168 MHz. C. 10.100 - 10.150 MHz.
B. 24.940 - 24.990 MHz. D. 29.500 - 29.700 MHz.

Repeater operation on high frequency is limited only to the top of the 10 meter band, between *29.5 MHz to 29.7 MHz*. No other high frequency (3-30MHz) bands permit repeater operation. [97.205] **ANSWER D.**

E1F12 Who may be the control operator of an auxiliary station?
A. Any licensed amateur operator.
B. Only Technician, General, Advanced or Amateur Extra Class operators.
C. Only General, Advanced or Amateur Extra Class operators.
D. Only Amateur Extra Class operators.

An amateur radio auxiliary station is a transceiver under your local control, which remotely controls another station, like a mountaintop repeater, or a distant high-ridge cross-band system, part of a cooperating amateur station network. *Any amateur radio operator, EXCEPT NOVICE CLASS, may set up an auxiliary station and be the control operator of that station.* But here is the BIG NEWS, as of 2007: Auxiliary operation, originally restricted to ham bands above 222 MHz, may now take place on the popular 2 meter band, above 144.50 MHz, except the satellite sub-band 145.8-146.0 MHz. The FCC noted that there was no apparent basis to conclude that allowing auxiliary stations to transmit on the 2 meter band would cause harmful interference to other stations' communications. This new ruling now makes dual band handheld transceivers, with a built-in TNC, take on some additional benefits of auxiliary operation, except for the Novice operator who may NOT be a control operator of an auxiliary station. (While Novice and Advanced Class licenses are no long being issued, there are hams who still hold valid Novice and Advanced Class licenses.) [97.201] **ANSWER B.**

E1F14 Under what circumstances might the FCC issue a "Special Temporary Authority" (STA) to an amateur station?
A. To provide for experimental amateur communications.
B. To allow regular operation on Land Mobile channels.
C. To provide additional spectrum for personal use.
D. To provide temporary operation while awaiting normal licensing.

Ham radio operators are always on the prowl for new types of emissions, and vacant frequencies where the *FCC may allow us special temporary authority*. While obtaining the FCC's STA requires diligence and numerous filings, ham operators have been successful in obtaining permission *to experiment with new emissions* not even found in the rules as well as operation on unused very low frequencies (VLF) for specific propagation studies. These experimental communications may lead to future rule makings on our behalf! [1.931] **ANSWER A.**

E1B09 Which amateur stations may be operated in RACES?
A. Only those club stations licensed to Amateur Extra class operators.
B. Any FCC-licensed amateur station except a Technician class operator's station.
C. Any FCC-licensed amateur station certified by the responsible civil defense organization for the area served.
D. Any FCC-licensed amateur station participating in the Military Affiliate Radio System (MARS).

All amateurs are eligible for RACES membership.
[97.407] **ANSWER C.**
☞ Visit: www.fema.gov/emergency/nims
www.emcomm.org/em

The RACES Logo

E1B10 What frequencies are authorized to an amateur station participating in RACES?
A. All amateur service frequencies authorized to the control operator.
B. Specific segments in the amateur service MF, HF, VHF and UHF bands.
C. Specific local government channels.
D. Military Affiliate Radio System (MARS) channels.

Anyone with a valid amateur service license, including a Novice Class license, is eligible to be certified by a civil defense organization as a RACES operator. However, *you do not gain any out-of-band privileges as a RACES operator.* [97.407]
ANSWER A. ☞ Visit: www.training.fema.gov/emiweb/is/crslist.asp

E1E03 What is a Volunteer Examiner Coordinator?
A. A person who has volunteered to administer amateur operator license examinations.
B. A person who has volunteered to prepare amateur operator license examinations.
C. An organization that has entered into an agreement with the FCC to coordinate amateur operator license examinations.
D. The person who has entered into an agreement with the FCC to be the VE session manager.

A Volunteer Examiner Coordinator is *an organization that has an agreement with the FCC to coordinate amateur radio license exams.* Two of the biggest VECs are the American Radio Relay League, and the W5YI-VEC. [97.521]
ANSWER C.

VE badge

E1E02 Where are the questions for all written U.S. amateur license examinations listed?

    A. In FCC Part 97.
    B. In a question pool maintained by the FCC.
    C. In a question pool maintained by all the VECs.
    D. In the appropriate FCC Report and Order.

The National Council of Volunteer Examiner Coordinators (NCVEC) has an internal Question Pool Committee that insures all exam questions are valid. Each of the three amateur radio question pools is updated every four years. *The question pools are maintained by all of the VECs.* Hams are encouraged to work with their VECs to help keep the pools stocked with fresh questions. The questions do not come from the FCC! [97.523] **ANSWER C.** ☞ **Visit: www.ncvec.org**

E1E04 Which of the following best describes the Volunteer Examiner accreditation process?

    A. Each General, Advanced and Amateur Extra Class operator is automatically accredited as a VE when the license is granted.
    B. The amateur operator applying must pass a VE examination administered by the FCC Enforcement Bureau.
    C. The prospective VE obtains accreditation from the FCC.
    D. The procedure by which a VEC confirms that the VE applicant meets FCC requirements to serve as an examiner.

VE accreditation is received from a Volunteer Examiner Coordinator *(VEC)* that *makes sure each VE applicant meets specific FCC requirements*, and has a clean record. [97.5091, 97.525] **ANSWER D.**

E1E01 What is the minimum number of qualified VEs required to administer an Element 4 amateur operator license examination?

    A. 5.
    B. 2.
    C. 4.
    D. 3.

It takes a *team of 3* Extra Class accredited examiners to administer an Element 4 amateur Extra Class test. It also takes these three Extra Class VEs to administer the General Class, Element 3, exam. [97.509] **ANSWER D.**

Working with kids as an Elmer or VE is both fun and a great way to help ham radio grow.

E1E08 To which of the following examinees may a VE not administer an examination?

    A. Employees of the VE.
    B. Friends of the VE.
    C. Relatives of the VE as listed in the FCC rules.
    D. All of these choices are correct.

As a volunteer examiner, *you may not test your own close relatives, friends, or fellow employees*. [97.509] **ANSWER C.**

E1E06 Who is responsible for the proper conduct and necessary supervision during an amateur operator license examination session?

    A. The VEC coordinating the session.
    B. The FCC.
    C. Each administering VE.
    D. The VE session manager.

Each one of the administering Volunteer Examiners must continuously supervise everyone in the room taking the exams. Administering VEs should not talk amongst themselves as the examinees are pouring over the test questions. [97.509] **ANSWER C.**

E1E10 What must the administering VEs do after the administration of a successful examination for an amateur operator license?

    A. They must collect and send the documents to the NCVEC for grading.
    B. They must collect and submit the documents to the coordinating VEC for grading.
    C. They must submit the application document to the coordinating VEC according to the coordinating VEC instructions.
    D. They must collect and send the documents to the FCC according to instructions.

If you are the head VE, also called the Administering VE, *you must submit the application documents to the coordinating VEC.* Follow the coordinating VEC instructions on how you actually send the proper paperwork to the VEC. [97.509(m)] **ANSWER C.**

E1E05 What is the minimum passing score on amateur operator license examinations?

    A. Minimum passing score of 70%.
    B. Minimum passing score of 74%.
    C. Minimum passing score of 80%.
    D. Minimum passing score of 77%.

As an accredited volunteer examiner, you need to know the number of questions an applicant might miss, but still achieving a *minimum passing grade of 74%.* Technician Class cannot miss more than 9 questions, and pass with 26 correct out of 35. General Class cannot miss more than 9 questions, and pass with 26 correct out of 35. Extra Class cannot miss more than 13 questions, and pass with 37 correct out of 50. [97.503(a)] **ANSWER B.**

E1E11 What must the VE team do if an examinee scores a passing grade on all examination elements needed for an upgrade or new license?

    A. Photocopy all examination documents and forward them to the FCC for processing.
    B. Three VEs must certify that the examinee is qualified for the license grant and that they have complied with the administering VE requirements.
    C. Issue the examinee the new or upgrade license.
    D. All these choices are correct .

When an examinee upgrades to a new, higher license, the 3-member VE team must *certify the exam paperwork and sign the CSCE* (Certificate of Successful Completion of Examination) showing that they have complied with the VE requirements. The applicant holds onto the CSCE until such time as the license is issued by the FCC. [97.509] **ANSWER B.**

Don't Lose Your CSCE – It Is Proof of Your Privileges Until Your License Arrives

E1E12 What must the VE team do with the application form if the examinee does not pass the exam?

A. Return the application document to the examinee.
B. Maintain the application form with the VEC's records.
C. Send the application form to the FCC and inform the FCC of the grade.
D. Destroy the application form.

You will *return the application to the examinee* and inform the candidate of the failing grade. Try to do this compassionately, urging the candidate to continue to study and try again soon. [97.509] **ANSWER A.**

E1E07 What should a VE do if a candidate fails to comply with the examiner's instructions during an amateur operator license examination?

A. Warn the candidate that continued failure to comply will result in termination of the examination.
B. Immediately terminate the candidate's examination.
C. Allow the candidate to complete the examination, but invalidate the results.
D. Immediately terminate everyone's examination and close the session.

If an examinee fails to comply with your instructions, you should *immediately terminate that candidate's examination.* [97.509] **ANSWER B.**

E1E14 For which types of out-of-pocket expenses do the Part 97 rules state that VEs and VECs may be reimbursed?

A. Preparing, processing, administering and coordinating an examination for an amateur radio license.
B. Teaching an amateur operator license examination preparation course.
C. No expenses are authorized for reimbursement.
D. Providing amateur operator license examination preparation training materials.

Volunteer examiners and VECs are allowed to recoup out-of-pocket *expenses for preparing, processing, administering, and coordinating an examination* for an amateur radio license. There is a set fee that each applicant pays for the entire exam session. [97.527] **ANSWER A.**

E1E13 What are the consequences of failing to appear for re-administration of an examination when so directed by the FCC?
  A. The licensee's license will be cancelled.
  B. The person may be fined or imprisoned.
  C. The licensee is disqualified from any future examination for an amateur operator license grant.
  D. All these choices are correct.

Occasionally, the FCC asks an applicant to appear for re-administration of an exam. This occurs if the exam session appears to have been flawed. If the licensee fails to appear for the re-exam, *the licensee's new license will be canceled* and a replacement license will be issued that is consistent with examination elements not invalidated. [97.519] **ANSWER A.**

E1E09 What may be the penalty for a VE who fraudulently administers or certifies an examination?
  A. Revocation of the VE's amateur station license grant and the suspension of the VE's amateur operator license grant.
  B. A fine of up to $1000 per occurrence.
  C. A sentence of up to one year in prison.
  D. All of these choices are correct.

Any VE who is caught administering a fraudulent exam may have his or her *station license revoked and the VE's amateur radio license grant suspended.* [97.509] **ANSWER A.**

E8D07 What is an electromagnetic wave?

A. Alternating currents in the core of an electromagnet.
B. A wave consisting of two electric fields at right angles to each other.
C. A wave consisting of an electric field and a magnetic field oscillating at right angles to each other.
D. A wave consisting of two magnetic fields at right angles to each other.

When you think of an *electromagnetic radio wave*, think of it as a *combination of an electric field and a magnetic field* at right angles to one another traveling in waves through the atmosphere at the same time. One field is vertically polarized and the other is horizontally polarized. If the antenna is vertically polarized, the electric field will be vertical, and the magnetic field will be horizontal. **ANSWER C.**

E8D08 Which of the following best describes electromagnetic waves traveling in free space?

A. Electric and magnetic fields become aligned as they travel.
B. The energy propagates through a medium with a high refractive index.
C. The waves are reflected by the ionosphere and return to their source.
D. Changing electric and magnetic fields propagate the energy .

Electromagnetic radio waves, traveling in free space, contain both an *electric field and a magnetic field propagating the energy*. The fields are at a right angle to each other, and it is both the electric field and the magnetic field that makes up the radio wave energy traveling at the speed of light in a vacuum. **ANSWER D.**

☞ **Visit: www.ips.gov.au/hf_systems/6/9/1**

The Effect of the Ionosphere on Radio Waves

To help you with the questions on radio wave propagation, here is a brief explanation on the effect the ionosphere has on radio waves.

The ionosphere is the electrified atmosphere from 40 miles to 400 miles above the Earth. You can sometimes see it as "northern lights." It is charged-up daily by the Sun, and does some miraculous things to radio waves that strike it. Some radio waves are absorbed during daylight hours by the ionosphere's D layer. Others are bounced back to Earth. Yet others penetrate the ionosphere and never come back again. The wavelength of the radio waves determines whether the waves will be absorbed, refracted, or will penetrate. Here's a quick way to memorize what the different layers do during day and nighttime hours:

The D layer is about 40 miles up. The D layer is a Daylight layer; it almost disappears at night. D for Daylight. The D layer absorbs radio waves between 1 MHz to 7 MHz. These are long wavelengths. All others pass through.

The E layer is also a daylight layer, and it is very Eccentric. E for Eccentric. Patches of E layer ionization may cause some surprising reflections of signals on both high frequency as well as very-high frequency. The E layer height is usually 70 miles.

The F1 layer is one of the layers farthest away. The F layer gives us those Far away signals. F for Far away. The F1 layer is present during daylight hours, and is up around 150 miles. The F2 layer is also present during daylight hours, and it gives us the Furthest range. The F2 layer is 250 miles high, and it's the best for the Farthest range on medium and short waves. The F2 layer is strongest in the summer months. During winter months, both the F1 and F2 layers may become unpredictable, but always strong enough to support exciting skywaves! At nighttime, the F1 and F2 layers combine to become just the F layer at 180 miles. This F layer at nighttime will usually bend radio waves between 1 MHz and 15 MHz back to earth. At night, the D and E layers disappear.

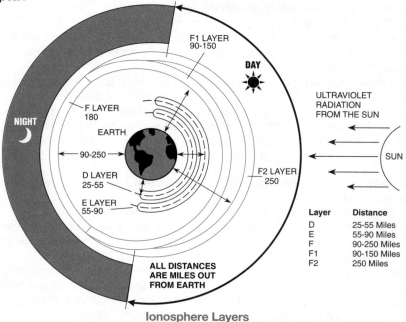

Ionosphere Layers

Source: *Antennas — Selection and Installation*, © 1986, Master Publishing, Inc., Niles, Illinois

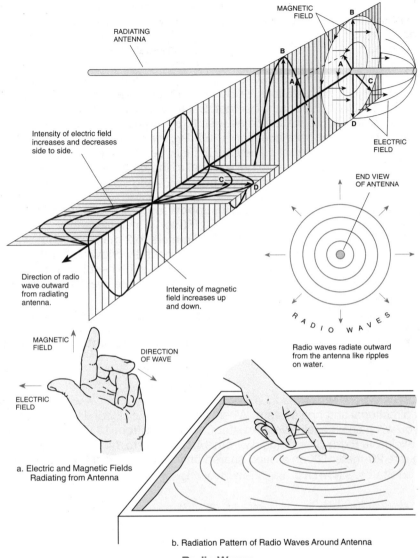

Intensity of electric field increases and decreases side to side.

Direction of radio wave outward from radiating antenna.

Intensity of magnetic field increases up and down.

END VIEW OF ANTENNA

Radio waves radiate outward from the antenna like ripples on water.

a. Electric and Magnetic Fields Radiating from Antenna

b. Radiation Pattern of Radio Waves Around Antenna

Radio Waves

Source: Basic Electronics © 1994, 2000 Master Publishing, Inc., Niles, Illinois

E8D09 What is meant by circularly polarized electromagnetic waves?

A. Waves with an electric field bent into a circular shape.

B. Waves with a rotating electric field.

C. Waves that circle the Earth.

D. Waves produced by a loop antenna.

Some satellite enthusiasts will use circular polarization to improve their transmission and reception through an orbiting satellite. With *circular polarization, the electric field rotates*. It could be right-hand circular, or left-hand circular, depending on the antenna feed system. **ANSWER B.**

E3B04 What type of propagation is probably occurring if an HF beam antenna must be pointed in a direction 180 degrees away from a station to receive the strongest signals?

A. Long-path.   C. Transequatorial.
B. Sporadic-E.   D. Auroral.

If two stations wish to communicate on high frequency when the sun is ionizing the F1 and F2 layers, both stations would normally aim their antennas at each other for an excellent "short path" contact. But if the sun is on the other side of the Earth to both stations, the stations may wish to try *long-path propagation* by aiming their antennas in the opposite direction – 1*80 degrees away from the station with which they wish to communicate*. They are actually sending their signals all the way around the World, the long way, to get a signal to the other station. Many times you can hear a tell-tale echo when a station is being received via a long path. **ANSWER A.**

E3B05 Which amateur bands typically support long-path propagation?

A. 160 to 40 meters.   C. 160 to 10 meters.
B. 30 to 10 meters.   D. 6 meters to 2 meters.

Long-range, *long-path* propagation can be found on the worldwide bands from *10 meters down to 160 meters*. Twenty meters is always the best band to expect almost daily long-path propagation, which is best in the late evening or early morning hours. **ANSWER C.**

☞ Visit: www.arrl.org/tis/info/propagation.html

E3B06 Which of the following amateur bands most frequently provides long-path propagation?

A. 80 meters.   C. 10 meters.
B. 20 meters.   D. 6 meters.

The amateur radio *20-meter band is the best one for long-path* DX with a relatively modest directional antenna system. When you pass your upcoming Extra Class exam, you will gain an exclusive 20-meter CW window from 14.000-14.025 MHz, and an exclusive Extra Class window from 14.150 MHz to 14.175 MHz, just for working rare 20-meter DX on a band shared only by other Extra Class operators in the United States. **ANSWER B.** ☞ Visit: www.ips.gov.au/hf_systems/6/9/1

	14.000 MHz		14.175 MHz		14.350 MHz
20 METERS	EX CW	CW RTTY DATA	EX VOICE	VOICE	
	14.025 MHz		14.150 MHz		

E3B07 Which of the following could account for hearing an echo on the received signal of a distant station?

A. High D layer absorption.
B. Meteor scatter.
C. Transmit frequency is higher than the MUF.
D. Receipt of a signal by more than one path.

When long-path conditions exist, European ham stations and even European shortwave stations will be received with a noticeable echo on the signal. The *echo is the slight delay of the signal coming in long-path following what you are also hearing sooner as a short-path signal*. **ANSWER D.**

☞ Visit: http://tf.nist.gov/stations/wwv.html

E3C08 What is the name of the high-angle wave in HF propagation that travels for some distance within the F2 region?
A. Oblique-angle ray.
B. Pedersen ray.
C. Ordinary ray.
D. Heaviside ray.

The Pedersen ray is a very-high-angle signal path occurring in the F-layer region. When band conditions are quiet it provides a strong first hop between Northeastern states and all of Europe. Best signal strengths occur when it is afternoon in Europe and early morning here in the U.S., with excitement from 20 meters all the way up through six meters! If you could visualize the *Pedersen ray*, it almost looks like it uses the F-layer as a waveguide on HF! **ANSWER B.**

E2C12 What might help to restore contact when DX signals become too weak to copy across an entire HF band a few hours after sunset?
A. Switch to a higher frequency HF band.
B. Switch to a lower frequency HF band.
C. Wait 90 minutes or so for the signal degradation to pass.
D. Wait 24 hours before attempting another communication on the band.

Now that we are cresting on the peak of solar cycle 24, many bands like 15 and 20 meters may remain "open" for skip signals well into the evening hours. When the band finally gives up and goes into noise with no more DX heard in the background, *switch to a lower frequency HF band* to see if that band is still open for skywave excitement. **ANSWER B.** ☞ **Visit: http://propagation.hfradio.org**

E3B08 What type of HF propagation is probably occurring if radio signals travel along the terminator between daylight and darkness?
A. Transequatorial.
B. Sporadic-E.
C. Long-path.
D. Gray-line.

Once you earn your Extra Class license, you will stay up for long hours working rare DX. Just as the sun is rising here, it may be setting over Europe or Asia, and working stations in the twilight at both ends of the circuit may lead to some extraordinary DX contacts that can last up to 30 minutes. This is called *gray-line propagation*. **ANSWER D.**

E3B10 What is the cause of gray-line propagation?
A. At midday, the Sun being directly overhead superheats the ionosphere causing increased refraction of radio waves.
B. At twilight, D-layer absorption drops while E-layer and F-layer propagation remain strong.
C. In darkness, solar absorption drops greatly while atmospheric ionization remains steady.
D. At mid afternoon, the Sun heats the ionosphere decreasing radio wave refraction and the MUF.

During twilight, the D layer quickly disappears resulting in less absorption, *while the E and F layers continue relatively strong*. This same condition holds true during solar eclipses. **ANSWER B.**

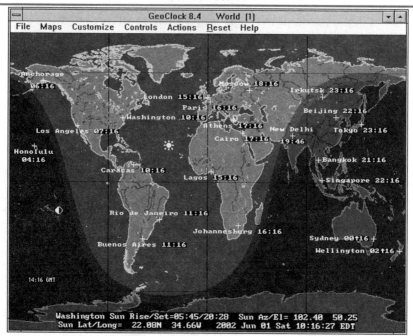

A Geo Clock shows the gray-line area.
http://www.mygeoclock.com/geoclock

E3B11 Which of the following describes gray-line propagation?
A. Backscatter contacts on the 10 meter band.
B. Over the horizon propagation on the 6 and 2 meter bands.
C. Long distance communications at twilight on frequencies less than 15 MHz.
D. Tropospheric propagation on the 2 meter and 70 centimeter bands.

Gray-line propagation occurs at sunrise and sunset when ultraviolet radiation is not ionizing the absorbing D-layer. The E-layer and F-layer are still receiving sunset and sunrise ionization, so our *radio signals below 15 MHz scoot along this area of twilight, over some extraordinary distances!* Gray-line propagation between two distant twilight zones (sunrise in California and sunset in Japan) may only last about half an hour, so get some rare DX with extraordinarily strong signals, for just a few minutes in the early morning and a few minutes at sundown. **ANSWER C.**

E3B09 At what time of day is gray-line propagation most likely to occur?
A. At sunrise and sunset.
B. When the Sun is directly above the location of the transmitting station.
C. When the Sun is directly overhead at the middle of the communications path between the two stations.
D. When the Sun is directly above the location of the receiving station.

Gray-line propagation occurs when the Sun is just setting to our west, and just coming up nearly halfway around the world to their east. Our signal travels along the gray-line "terminator" and provides about twenty minutes of exciting stronger-than-usual propagation. There are numerous computer programs that show gray-line DX opportunities. If you enjoy working DX just before *sunrise*, turn your antennas to the east and maybe work a station that is just enjoying their *sunset*. The twilight contacts can really be a thrill! **ANSWER A.**

E3A08 When a meteor strikes the Earth's atmosphere, a cylindrical region of free electrons is formed at what layer of the ionosphere?

A. The E layer.
B. The F1 layer.
C. The F2 layer.
D. The D layer.

It is within the *E-layer* of the ionosphere that 6- and 2-meter excitement like Sporadic-E and *meteor scatter contacts* take place. Think mEtEor! **ANSWER A.**

E3A09 Which of the following frequency ranges is well suited for meteor-scatter communications?

A. 1.8 - 1.9 MHz.
B. 10 - 14 MHz.
C. 28 - 148 MHz.
D. 220 - 450 MHz.

During a recent meteor shower, I was able to maintain 30-second meteor-scatter QSOs on the 10-meter band; 10-second QSOs on the 6-meter band, and 3-second QSOs on the 2-meter band. The higher you go in frequency, the shorter the contact time. *28-148 MHz are the best* areas to hear the effects of *meteor-scatter*. If you live within 100 miles of an airport, listen to 75 MHz on a scanner for airport radio navigation signals to bounce off of meteor trails. **ANSWER C.**

E2D01 Which of the following digital modes is especially designed for use for meteor scatter signals?

A. WSPR.
B. FSK441.
C. Hellschreiber.
D. APRS.

If you are into computers, ham radio digital signaling is just the thing for you! There are many software programs – most free – that may work with your computer's sound card; and there are excellent add-on radio signal controllers that let the more sophisticated transceivers both transmit and receive ham radio digital calls. Up on VHF and UHF, the most popular digital mode for *meteor scatter* contacts is *FSK 441 (Frequency Shift Keying)*. **ANSWER B.**

☞ Visit: http://spaceweather.com

E3A10 Which of the following is a good technique for making meteor-scatter contacts?

A. 15 second timed transmission sequences with stations alternating based on location.
B. Use of high speed CW or digital modes.
C. Short transmission with rapidly repeated call signs and signal reports.
D. All of these choices are correct.

There are several different ways of conducting *meteor scatter contacts*. If you and your friend 500 miles away want to set a meteor scatter schedule, you first make sure your clocks are perfectly in sync with WWV at 10,000 kHz. Agree on who takes the first and third *15 second calls*, and then each listen for the other's fifteen second transmission. Transmitting west to east takes the first and third 15 second period. Another way for random meteor scatter contacts is through *high speed CW, digital modes, or rapidly giving your call sign for a few seconds*. Listen to my audio CDs to hear exciting meteor scatter contacts. **ANSWER D.**

☞ Visit: http://pulsar.princeton.edu/~joe/K1JT
www.VHFDX.DE/WSJT

E3B01 What is transequatorial propagation?
    A. Propagation between two mid-latitude points at approximately the same
       distance north and south of the magnetic equator.
    B. Propagation between any two points located on the magnetic equator.
    C. Propagation between two continents by way of ducts along the magnetic
       equator.
    D. Propagation between two stations at the same latitude.
A TE (transequatorial) signal is propagated between stations north and south of
the magnetic Equator. Some of the best contacts have been between Florida and
countries in South America. Any *stations* in contact with each other are *about the
same distance away from the Equator* – ideal conditions for a TE contact. Both CW
as well as SSB are the normal modes for TE on VHF and UHF bands. **ANSWER A.**

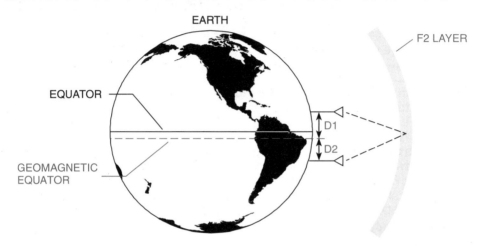

For transequatorial propagation to occur, stations at D1 and D2 must be
equidistant from the geomagnetic equator. Popular theory is that reflections
occur in and off the F2 layer during peak sunspot activity.

Transequatorial Propagation

**E3B02 What is the approximate maximum range for signals using
transequatorial propagation?**
    A. 1000 miles.   C. 5000 miles.
    B. 2500 miles.   D. 7500 miles.
The maximum total distance for a TE (transequatorial) contact is about *5,000 miles.*
ANSWER C.

E3B03 What is the best time of day for transequatorial propagation?
    A. Morning.   C. Afternoon or early evening.
    B. Noon.   D. Late at night.
Similar to Sporadic-E, the *best time of day for TE* propagation is *mid-afternoon or
early evening*. By this time of day, the Sun's ultraviolet radiation has had time to
"warm up" the possible path. **ANSWER C.**

E3C02 What is the cause of Aurora activity?
A. The interaction between the solar wind and the Van Allen belt.
B. A low sunspot level combined with tropospheric ducting.
C. The interaction of charged particles from the Sun with the Earth's magnetic field and the ionosphere.
D. Meteor showers concentrated in the northern latitudes.

During periods of *major sunspots*, auroral activity will be high. When a giant solar flare occurs on the face of the Sun, we see it here on earth about eight minutes later and feel the effects of its ultraviolet radiation. About two days later, slower moving *charged particles* begin to pump up the ionosphere, many times *creating spectacular auroral light shows*. It's the charged particles that take a couple of days to finally reach us that give us an aurora.
ANSWER C. ☞ **Visit: www.aurorasentry.net**

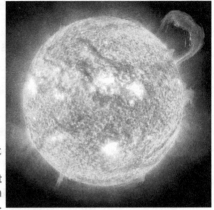

Solar flares and sunspots affect radiowave propagation
Photo courtesy of N.A.S.A.

E3C03 Where in the ionosphere does Aurora activity occur?
A. In the F1-region.   C. In the D-region.
B. In the F2-region.   D. In the E-region.
Have you been to Canada or Alaska to watch an *aurora*? These light shows *occur in the "E" layer* of our ionosphere. On worldwide signals, the aurora will many times degrade good skip conditions. But on VHF and UHF, the aurora may provide some exciting, fluttery "E" layer bounces. **ANSWER D.**
☞ **Visit: http://umtof.umd.edu/pm/latest2day.gif**

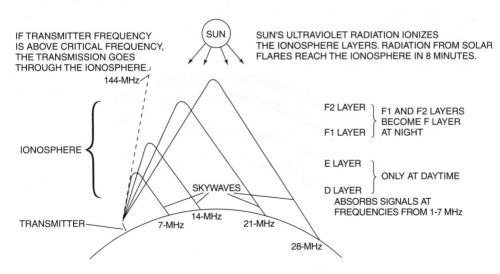

Ionosphere and Its Layers

E3C01 Which of the following effects does Aurora activity have on radio communications?

A. SSB signals are raspy.
B. Signals propagating through the Aurora are fluttery.
C. CW signals appear to be modulated by white noise.
D. All of these choices are correct.

During periods of stormy ionospheric action, try some aurora contacts. You may first notice that *CW signals have almost no tone*, but you hear them as bursts of white noise dots and dashes. *Single sideband signals will sound raspy and fluttery*, and you can hear all of these sounds on my audio CD course. Take a listen!
ANSWER D. ☞ Visit: www.csvhfs.org

E3C04 Which emission mode is best for Aurora propagation?

A. CW.
B. SSB.
C. FM.
D. RTTY.

Bouncing a signal off of an aurora is difficult. CW always provides the best opportunity for someone hearing raspy sounds coming into their receiver. On many *auroral contacts on CW*, we don't even hear a tone – all that can be heard are just short and long rushes of super-imposed noise. Play my audio CD Extra Class course and listen for yourself. **ANSWER A.**

E3C11 From the contiguous 48 states, in which approximate direction should an antenna be pointed to take maximum advantage of aurora propagation?

A. South.
B. North.
C. East.
D. West.

You would *aim your directional antenna North* at the aurora in order to pick up VHF and UHF fluttery auroral CW sounds. Also, with big auroras, you can sometimes even transmit SSB. Forget about FM or RTTY, these modes usually will not work off of an aurora. **ANSWER B.**

Geomagnetic disturbances caused by the Sun result in the Northern Lights.

E3C05 Which of the following describes selective fading?

A. Variability of signal strength with beam heading.
B. Partial cancellation of some frequencies within the received pass band.
C. Sideband inversion within the ionosphere.
D. Degradation of signal strength due to backscatter.

Selective fading occurs when incoming signals take slightly different paths, with the signal slowly rising and falling in signal strength like waves on a sea shore. Up at 29.6 MHz FM, your extra-wide signal will regularly experience selective fading, *cancelling some of the voice frequencies within the received pass band*. You can hear an example on my audio course. Have a listen! **ANSWER B.**

E4E06 What is a major cause of atmospheric static?
    A. Solar radio frequency emissions. C. Geomagnetic storms.
    B. Thunderstorms.   D. Meteor showers.
Thunderstorms carry lightning activity, and lightning noise can be received from as far away as 800 miles on the 80- and 40-meter bands. **ANSWER B.**

E3C14 Why does the radio-path horizon distance exceed the geometric horizon?
    A. E-region skip.
    B. D-region skip.
    C. Downward bending due to aurora refraction.
    D. Downward bending due to density variations in the atmosphere.
The VHF/UHF radio horizon is approximately 15% farther away than the geometric horizon. This is because the *lower portion of the VHF/UHF wave front is slightly slowed by our denser atmosphere*, allowing the *higher altitude wave front to travel slightly faster*, causing the entire wave front to "bend" slightly beyond the geometric horizon. Listen to my exclusive audio CD course for the sound of long range VHF/UHF over-the-horizon range! **ANSWER D.**

VHF Range Nomograph

E3C06 By how much does the VHF/UHF radio-path horizon distance exceed the geometric horizon?
    A. By approximately 15% of the distance.
    B. By approximately twice the distance.
    C. By approximately one-half the distance.
    D. By approximately four times the distance.
Radio waves bend slightly over the horizon because of the difference in the air's refractive index at higher altitudes. You can calculate your approximate VHF range to the horizon with the formula: D (miles) = $1.415 \times \sqrt{2H}$ (in feet). Determine the height (in feet) of the station you are wishing to communicate with, calculate its distance to the horizon, and add the two answers together for your rock-solid communications range. Depending on local weather conditions, a *15% to 30% range enhancement over the geometric horizon* will usually take place at VHF and UHF radio frequencies. **ANSWER A.**

E3C09 **Which of the following is usually responsible for causing VHF signals to propagate for hundreds of miles?**
    A. D-region absorption.   C. Tropospheric ducting.
    B. Faraday rotation.   D. Ground wave.
Some incredible extra-long-range VHF and UHF propagation may take place in the presence of a high-pressure system. Sinking warm air, called a subsidence, overlays cool air just above the surface of the Earth and over water. A well-defined inversion layer occurs when there is a sharp boundary between the sinking warm air and the cool air beneath. Over big cities, we call it smog. To the VHF and UHF DX enthusiast, we call it a for-sure key to super range contacts via VHF tropospheric ducting. I've communicated many times from my station in California to Hawaii on the 2-meter band during summer months when *tropospheric ducting* is at its best! **ANSWER C.** ☞ **Visit: http://dx.qsl.net/propagation/tropo.php**

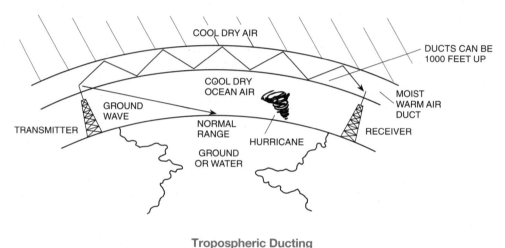

Tropospheric Ducting

E3C12 **How does the maximum distance of ground-wave propagation change when the signal frequency is increased?**
    A. It stays the same.   C. It decreases.
    B. It increases.   D. It peaks at roughly 14 MHz.
Ground waves travel further on lower frequencies, so as the *frequency* of a signal is *increased, ground wave propagation decreases*. This same propagation phenomena occurs with audio frequencies. Lower audio frequencies travel further than higher audio frequencies. That's why fog horns always blast out a very low audio tone. **ANSWER C.**

E3C13 **What type of polarization is best for ground-wave propagation?**
    A. Vertical.   C. Circular.
    B. Horizontal.   D. Elliptical.
We use *vertical polarization for best ground-wave propagation*. Knife-edge bending is a good example of vertically-polarized ground waves that drag the bottom half of the wave over a mountain top, resulting in the ground-wave signal getting into valleys where normally there would be no reception. **ANSWER A.**

RADIO ENERGY DIFFRACTED

RECEIVER

TRANSMITTER

MOUNTAIN

Knife-Edge Diffraction

E2C01 Which of the following is true about contest operating?
   A. Operators are permitted to make contacts even if they do not submit a log.
   B. Interference to other amateurs is unavoidable and therefore acceptable.
   C. It is mandatory to transmit the call sign of the station being worked as part of every transmission to that station.
   D. Every contest requires a signal report in the exchange.

Many of you are preparing for Extra Class to gain valuable, exclusive Extra Class frequencies for contesting. Contesting hones our capability to pick weak signals out of the noise, and enhances speedy operating techniques, essential in some emergency communication situations. *There is no requirement to submit a contest log* when participating in a contest. However, sending in a log lets contest officials know the popularity of their specific contest. **ANSWER A.**

E2C11 How should you generally identify your station when attempting to contact a DX station working a pileup or in a contest?
   A. Send your full call sign once or twice.
   B. Send only the last two letters of your call sign until you make contact.
   C. Send your full call sign and grid square.
   D. Send the call sign of the DX station three times, the words this is, then your call sign three times.

Always make sure to *send your full call sign* when operating DX, not just the last letter or 2 in your call sign, which is technically not legal. Full call sign. Rules also require that you momentarily listen to your offset transmit frequency to ensure the frequency is not in use. **ANSWER A.**

E2C10 Why might a DX station state that they are listening on another frequency?
   A. Because the DX station may be transmitting on a frequency that is prohibited to some responding stations.
   B. To separate the calling stations from the DX station.
   C. To reduce interference, thereby improving operating efficiency.
   D. All of these choices are correct.

As a new Extra Class operator, learn from listening during your first contest in the Extra Class portion of the band. If it is an international contest, you will hear rare DX stations calling for a contact "listening up 10." This means they are not listening on their own transmit frequency, but are listening 10 kHz higher for a response.

An example might be a foreign DX station operating just below your Extra Class band limits on 20 meters, at 14.145 MHz, listening "up 10." This means you would transmit back to them on 14.155 MHz. To do this, you tune VFO-A to the receive frequency of 14.145 MHz, and put VFO-B on your transmit frequency of 14.155 MHz. Now SPLIT VFO OPERATION and make sure you are transmitting on VFO-B, yet receiving on VFO-A. You must be in SPLIT for this to occur. If you transmit back to the foreign station outside of your Extra Class lower band edge you are illegally out of band and the foreign station can't hear you because they are listening "up 10 kHz" within the U.S. Extra Class band. *Split operation reduces interference, separates calling stations from the DX station, and lets the DX station stay in the clear, out of our U.S. phone band.* **ANSWER D.**

☞ **Visit: www.arrl.org/contests**

E2C07 What is the Cabrillo format?
 A. A standard for submission of electronic contest logs.
 B. A method of exchanging information during a contest QSO.
 C. The most common set of contest rules.
 D. The rules of order for meetings between contest sponsors.

The various computer logging programs (such as CT by K1EA, TR Log, N1MM, etc.) have built-in utilities that *automatically generate a Cabrillo-formatted log* after the contest is over. It's one of those features where you click on a box and the program automatically ker-chunks out a Cabrillo file for submission to the ARRL, CQ, or other contest sponsor. **ANSWER A.**

E2C02 Which of the following best describes the term "self-spotting" in regards to contest operating?
 A. The generally prohibited practice of posting one's own call sign and frequency on a call sign spotting network.
 B. The acceptable practice of manually posting the call signs of stations on a call sign spotting network.
 C. A manual technique for rapidly zero beating or tuning to a station's frequency before calling that station.
 D. An automatic method for rapidly zero beating or tuning to a station's frequency before calling that station.

Most avid contesters, looking for a high contest score, will stay glued to a computer call sign spotting network. This allows them to see rare stations and where they are operating on the radio dial. It is considered *poor practice to spot yourself on a computer posting* of contest activity, called "self spotting." **ANSWER A.**

E2C04 On which of the following frequencies is an amateur radio contest contact generally discouraged?
 A. 3.525 MHz.
 B. 14.020 MHz.
 C. 28.330 MHz.
 D. 146.520 MHz.

On the 2 meter band, *146.520 MHz is the national simplex calling frequency*. Most VHF contests preclude any operation on this important calling channel, so don't use "52" for contesting. **ANSWER D.**

☞ **Visit: www.hornucopia.com/contestcal/index.html**

E2C06 During a VHF/UHF contest, in which band segment would you expect to find the highest level of activity?
    A. At the top of each band, usually in a segment reserved for contests.
    B. In the middle of each band, usually on the national calling frequency.
    C. In the weak signal segment of the band, with most of the activity near the calling frequency.
    D. In the middle of the band, usually 25 kHz above the national calling frequency.

During a VHF/UHF contest, *most of the activity is found on each side of the calling frequency*. **ANSWER C.** ☞ **Visit: www.swotrc.org**

Band	Calling Frequency
6 meters	50.125 MHz
2 meters	144.200 MHz
1.25	222.1 MHz
70 cm	432.100 MHz
35 cm	902.100 MHz
23 cm	1296.100 MHz

E2C03 From which of the following bands is amateur radio contesting generally excluded?
    A. 30 meters.   C. 2 meters.
    B. 6 meters.   D. 33 cm.

The 30 meter, 10 MHz band allows CW and data only with a maximum power of 200 watts output. *There is no contesting allowed on the 30 meter band*. **ANSWER A.**

E2C05 What is the function of a DX QSL Manager?
    A. To allocate frequencies for DXpeditions.
    B. To handle the receiving and sending of confirmation cards for a DX station.
    C. To run a net to allow many stations to contact a rare DX station.
    D. To relay calls to and from a DX station.

It's fun going out on a DXpedition. When I was at Christmas Island, operating T32GW, the QSL cards poured in for months! Sometimes, on a big scale DXpedition, minding the paper QSL cards and the electronic QSL requests is best handled by a *DX QSL manager*. They carefully *check the logs to insure all of the contact information is valid, and then send out the coveted QSL card confirmation*. It is a huge job, so if you know of a DX QSL manager, give them a good pat on the back! **ANSWER B.**

QSL Cards.

HERE ARE SOME WEB ADDRESSES OF USEFUL SITES RELATED TO SKYWAVES & CONTESTING:

http://www.ncdxf.org/beacon/beaconprograms/html
http://www.dxatlas.com/
http://www.mygeoclock.com/acehf/
http://dx-world.net/
http://www.voacap.com/prediction.html
http://hamradionation.com

Outer Space Comms

CQ EARTH...
CQ EARTH...

E1D02 What is the amateur satellite service?
   A. A radio navigation service using satellites for the purpose of self training, intercommunication and technical studies carried out by amateurs.
   B. A spacecraft launching service for amateur-built satellites.
   C. A radio communications service using amateur radio stations on satellites.
   D. A radio communications service using stations on Earth satellites for public service broadcast.

The amateur satellite service allows hams to *communicate through amateur station repeaters and translators located on orbiting satellites*. [97.3] **ANSWER C.**

E2A02 What is the direction of a descending pass for an amateur satellite?
   A. From north to south.   C. From east to west.
   B. From west to east.   D. From south to north.

The satellite will pass over the Equator going *from* our *Northern* Hemisphere *into* the *Southern* Hemisphere. **ANSWER A.**

E2A01 What is the direction of an ascending pass for an amateur satellite?
   A. From west to east.   C. From south to north.
   B. From east to west.   D. From north to south.

Ground tracking on an ascending pass will have the ham satellite going over the Equator *from* the *Southern* Hemisphere *to* the *Northern* Hemisphere. **ANSWER C.**

E2A03 What is the orbital period of an Earth satellite?
A. The point of maximum height of a satellite's orbit.
B. The point of minimum height of a satellite's orbit.
C. The time it takes for a satellite to complete one revolution around the Earth.
D. The time it takes for a satellite to travel from perigee to apogee.
One complete orbit is the period of a satellite. **ANSWER C.**

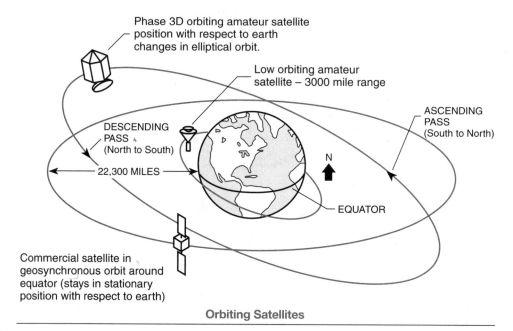

Phase 3D orbiting amateur satellite position with respect to earth changes in elliptical orbit.

Low orbiting amateur satellite – 3000 mile range

ASCENDING PASS (South to North)

DESCENDING PASS (North to South)

22,300 MILES

N

EQUATOR

Commercial satellite in geosynchronous orbit around equator (stays in stationary position with respect to earth)

Orbiting Satellites

E1D04 What is an Earth station in the amateur satellite service?
A. An amateur station within 50 km of the Earth's surface intended for communications with amateur stations by means of objects in space.
B. An amateur station that is not able to communicate using amateur satellites.
C. An amateur station that transmits telemetry consisting of measurement of upper atmosphere data from space.
D. Any amateur station on the surface of the Earth.
Below 50 kilometers, you're considered an *Earth station* when you are communicating via the amateur satellite service. [97.3] **ANSWER A.**

E1D11 Which amateur stations are eligible to operate as Earth stations?
A. Any amateur station whose licensee has filed a pre-space notification with the FCC's International Bureau.
B. Only those of General, Advanced or Amateur Extra Class operators.
C. Only those of Amateur Extra Class operators.
D. Any amateur station, subject to the privileges of the class of operator license held by the control operator.
An amateur Earth station is located down here on planet Earth. Operating as an *Earth station* for space communications *is available to all amateur stations*. Be sure to use only those frequency privileges allowed by your license class. [97.209] **ANSWER D.** ☞ **Visit: www.amsat.org**

E1D03 What is a telecommand station in the amateur satellite service?
- A. An amateur station located on the Earth's surface for communications with other Earth stations by means of Earth satellites.
- B. An amateur station that transmits communications to initiate, modify or terminate functions of a space station.
- C. An amateur station located more than 50 km above the Earth's surface.
- D. An amateur station that transmits telemetry consisting of measurements of upper atmosphere data from space.

Telecommand means *initiating, modifying, or terminating a function of another station*, such as a space station. [97.3] **ANSWER B.**

E1D10 Which amateur stations are eligible to be telecommand stations?
- A. Any amateur station designated by NASA.
- B. Any amateur station so designated by the space station licensee, subject to the privileges of the class of operator license held by the control operator.
- C. Any amateur station so designated by the ITU.
- D. All of these choices are correct.

Any amateur station so designated by the space station licensee may be a telecommand station. [97.211] **ANSWER B.**

E1D01 What is the definition of the term telemetry?
- A. One-way transmission of measurements at a distance from the measuring instrument.
- B. Two-way radiotelephone transmissions in excess of 1000 feet.
- C. Two-way single channel transmissions of data.
- D. One-way transmission that initiates, modifies, or terminates the functions of a device at a distance.

We hear the word "telemetry" a lot in satellite service. Satellite telemetry may give ground controllers vital information on satellite battery power, power loads, and satellite solar power charging. *Telemetry is a one-way transmission of measurements at a distance* from the measuring instrument. [97.3] **ANSWER A.**

E2A04 What is meant by the term mode as applied to an amateur radio satellite?
- A. The type of signals that can be relayed through the satellite.
- B. The satellite's uplink and downlink frequency bands.
- C. The satellite's orientation with respect to the Earth.
- D. Whether the satellite is in a polar or equatorial orbit.

Working through amateur satellites is exciting, challenging, and fun! Our transmit and receive frequencies for satellite operation are on separate bands. Each band has its own designator. The satellite receives on one frequency band, and transmits on another frequency band. Since two bands are involved, we will have two *mode designators* to best describe our transmit and receive bands, and the satellite receive and transmit bands. We call our Earth transmit band an *"uplink"*, and our receive band the *"downlink"*. **ANSWER B.**

E2A05 What do the letters in a satellite's mode designator specify?
- A. Power limits for uplink and downlink transmissions.
- B. The location of the ground control station.
- C. The polarization of uplink and downlink signals.
- D. The uplink and downlink frequency ranges.

The two *mode designators indicate* our transmit *uplink frequency* and our receive *downlink frequency* band. **ANSWER D.**

E2A09 What do the terms L band and S band specify with regard to satellite communications?
   A. The 23 centimeter and 13 centimeter bands.
   B. The 2 meter and 70 centimeter bands.
   C. FM and Digital Store-and-Forward systems.
   D. Which sideband to use.

You can work this out by recalling the MHz to meters and meters to MHz formula, but there are lots of conversions because we are working with centimeters and GHz. The L band at 1.26 GHz works out to be 23 centimeters, and the S band, at 2.4 GHz, works out to be 13 centimeters. You can rule out Answer B as that would be VHF and UHF, with C and D having nothing to do with band specifics. *23 centimeters is L band, and 13 centimeters is S band.* **ANSWER A.**

Frequency Bands	Frequency Range	Modes
High Frequency	21 - 30 MHz	Mode H
VHF	144 - 146 MHz	Mode V
UHF	435 - 438 MHz	Mode U
L band	1.26 - 1.27 GHz	Mode L
S band	2.4 - 2.45 GHz	Mode S
C band	5.8 GHz	Mode C
X band	10.4 GHz	Mode X
K band	24 GHz	Mode K

E2A06 On what band would a satellite receive signals if it were operating in mode U/V?
   A. 435-438 MHz.   C. 50.0-50.2 MHz.
   B. 144-146 MHz.   D. 29.5 to 29.7 MHz.

Watch out on this question – I don't want you to miss it even though you thought they might be trying to trick you. The question asks on what band would a satellite receive signals – this is the same as asking on what band would an Earth station transmit to the satellite. Since the *mode* begins with the letter *U*, we transmit, and the satellite simultaneously receives, on *UHF 435 - 438 MHz*, 70 cm, sometimes called the *"432" band.* **ANSWER A.**

E2A07 Which of the following types of signals can be relayed through a linear transponder?
   A. FM and CW.   C. PSK and Packet.
   B. SSB and SSTV.   D. All of these choices are correct.

Thanks to AMSAT, many of our OSCAR satellites offer a "bent pipe" re-transmission of many modes of our uplink signal. If we uplink FM and CW, it comes back down FM and CW through the satellite linear transponder. Same thing with single sideband, slow scan television, and the digital modes – *all of these signals are relayed through* the satellite's on-board *linear transponder.* **ANSWER D.**

E2A08 Why should effective radiated power to a satellite which uses a linear transponder be limited?
    A. To prevent creating errors in the satellite telemetry.
    B. To avoid reducing the downlink power to all other users.
    C. To prevent the satellite from emitting out of band signals.
    D. To avoid interfering with terrestrial QSOs.
Our ham satellites are powered by on-board batteries charged by solar panels. On older satellites, *any strong input signal is re-transmitted strong, quickly depleting satellite battery power*. Our modern constellation of low Earth orbit satellites keep modest power re-transmitted signals down to approximately the same output power, conserving battery life. In fact, an overly strong input signal can cause that resulting output signal to actually decrease in power, so don't run any more power than it takes to hear your own signal coming back, in the clear. **ANSWER B.**

E2A10 Why may the received signal from an amateur satellite exhibit a rapidly repeating fading effect?
    A. Because the satellite is spinning.
    B. Because of ionospheric absorption.
    C. Because of the satellite's low orbital altitude.
    D. Because of the Doppler Effect.
Many amateur satellites continuously rotate to keep all sides of the "bird" equally bathed in sunlight. This *spinning* distributes the heat build-up, and allows all of the solar panels to receive energy from the Sun. Because of the rotation, we hear a *characteristic rapid pulse in and out on the downlink*. **ANSWER A.**

E2A11 What type of antenna can be used to minimize the effects of spin modulation and Faraday rotation?
    A. A linearly polarized antenna.
    B. A circularly polarized antenna.
    C. An isotropic antenna.
    D. A log-periodic dipole array.
The circular polarized antenna will give you best results when working amateur satellites. Although you lose approximately 3 dB gain over a horizontal antenna, *circular polarization* minimizes fading. **ANSWER B.**

Tracking and communicating through amateur satellites can be done with a cross-polarized or circularly-polarized satellite antenna.

E2A12 What is one way to predict the location of a satellite at a given time?
    A. By means of the Doppler data for the specified satellite.
    B. By subtracting the mean anomaly from the orbital inclination.
    C. By adding the mean anomaly to the orbital inclination.
    D. By calculations using the Keplerian elements for the specified satellite.

You can log onto many satellite web pages and learn the *Keplerian elements* for a specific satellite or a specific day and time *for the location of the satellite*. You can find this information on the AMSAT website at **www.AMSAT.org**. **ANSWER D.**

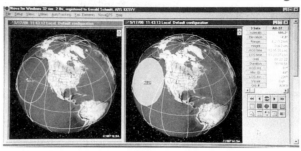

Computer programs and websites can show you where and when an amateur satellite or the Space Station will be in range of your ham station

Good to Know:

Where's the amateur satellite?

AMSAT is the North American distributor of SatPC32, a tracking program designed for ham satellite applications. It has a host of features that help you locate and track amateur satellites, set your antenna rotator for automatic control, and provides visual mapping of satellite locations as they orbit the Earth. You can download a demo version from www.dk1tb.de/indexeng.htm, or order online at www.amsat.org or by calling 1-888-322-6728. BTW – the author of SatPC32, DK1TB, donated the program to AMSAT and all proceeds support AMSAT.

E2D04 What is the purpose of digital store-and-forward functions on an Amateur Radio satellite?
   A. To upload operational software for the transponder.
   B. To delay download of telemetry between satellites.
   C. To store digital messages in the satellite for later download by other stations.
   D. To relay messages between satellites.
Amateur orbiting satellites make great store-and-forward devices to *hold messages that may be downloaded later* by other stations. **ANSWER C.**

E2D05 Which of the following techniques is normally used by low Earth orbiting digital satellites to relay messages around the world?
   A. Digipeating.   C. Multi-satellite relaying.
   B. Store-and-forward.   D. Node hopping.
Store-and-forward is a popular way of sending a wireless e-mail that may be received halfway around the world when the satellite passes over the distant station. **ANSWER B.**

E2A13 What type of satellite appears to stay in one position in the sky?
   A. HEO.   C. Geomagnetic.
   B. Geostationary.   D. LEO.

Great news from AMSAT – they are in discussion with Intelsat to place an amateur radio payload on a future Intelsat *geostationary* satellite in a "rideshare" arrangement. We could end up with a geostationary satellite that could provide 500 voice channels and the ability to provide video and other digital services, 24 hours a day, with dishes for S band and C band aimed in just *one fixed direction*! We currently have no ham satellites in geostationary orbit, but with this "rideshare" agreement with Intellsat, a "geo" may be in our future! Watch out for incorrect answer "geomagnetic." **ANSWER B.**

E1D06 Which of the following special provisions must a space station incorporate in order to comply with space station requirements?
    A. The space station must be capable of terminating transmissions by
       telecommand when directed by the FCC.
    B. The space station must cease all transmissions after 5 years.
    C. The space station must be capable of changing its orbit whenever such a
       change is ordered by NASA.
    D. All of these choices are correct.
All space stations must be capable of receiving a *telecommand to shut them down* if ordered to do so by the FCC. [97.207] **ANSWER A.**

E1D07 Which amateur service HF bands have frequencies authorized to space stations?
    A. Only 40m, 20m, 17m, 15m, 12m and 10m.
    B. Only 40m, 20m, 17m, 15m and 10m bands.
    C. 40m, 30m, 20m, 15m, 12m and 10m bands.
    D. All HF bands.
By international agreement, all high-frequency ITU voice bands plus most of our regular high-frequency ham bands are available for space operation. This includes *40 meters, 20 meters, 17 meters, 15 meters, 12 meters, and 10 meters*. [97.207]
ANSWER A.

E1D05 What class of licensee is authorized to be the control operator of a space station?
    A. All except Technician Class.
    B. Only General, Advanced or Amateur Extra Class.
    C. All classes.
    D. Only Amateur Extra Class.
There is no longer a specific higher grade license requirement for space operation. *Any grade of license* now qualifies to be the control operator. (Originally, only an Extra Class operator could transmit from space.) [97.207] **ANSWER C.**

E1D08 Which VHF amateur service bands have frequencies available for space stations?
    A. 6 meters and 2 meters.
    B. 6 meters, 2 meters, and 1.25 meters.
    C. 2 meters and 1.25 meters.
    D. 2 meters.
On VHF (30-300 MHz), *only the 2-meter band allows space station operation*. Space station operation is prohibited on 1.25 meters (the 222 MHz band) and 6 meters. You can find the right answer by looking at the wrong answers. [97.207] **ANSWER D.**

E1D09 Which amateur service UHF bands have frequencies available for a space station?

A. 70 cm.
B. 70 cm, 23 cm, 13 cm.
C. 70 cm and 33 cm.
D. 33 cm and 13 cm.

On UHF (300-3000 MHz), the 33 cm (900 MHz) band is NOT allowed for space station, but all other bands are allowed. This includes *70 cm (435 MHz), 23 cm (1260 MHz), and 13 cm (2300 MHz)*. [97.207] **ANSWER B.**

E3A01 What is the approximate maximum separation measured along the surface of the Earth between two stations communicating by Moon bounce?

A. 500 miles, if the Moon is at perigee.
B. 2000 miles, if the Moon is at apogee.
C. 5000 miles, if the Moon is at perigee.
D. 12,000 miles, as long as both can "see" the Moon.

It takes a minimum of a pair of long-boom Yagis, plus the maximum legal power output, to establish a Moon bounce QSO on CW or SSB phone. If you are working a station with a huge dish antenna, you might get by with a single long-boom Yagi and a good VHF or UHF kilowatt amplifier. Moon bounce is most common on the bottom edge of the 2-meter band, and to communicate over any distance, *both stations must be able to see the Moon* in the sky. Monthly ham magazines publish the best EME (Earth-Moon-Earth) days. **ANSWER D.**

E3A04 What type of receiving system is desirable for EME communications?

A. Equipment with very wide bandwidth.
B. Equipment with very low dynamic range.
C. Equipment with very low gain.
D. Equipment with very low noise figures.

For EME communications, it's best to use a receiver with front-end transistors that have extremely high gain and *extremely low noise* (a "hot front end"). For my station, I use GaAsFET transistors for maximum gain and minimum noise. **ANSWER D.**

EME communications require specialized equipment and patience.

E3A06 What frequency range would you normally tune to find EME signals in the 2 meter band?

A. 144.000 - 144.001 MHz.

B. 144.000 - 144.100 MHz.

C. 144.100 - 144.300 MHz.

D. 145.000 - 145.100 MHz.

Moon bounce is found in the CW weak-signal portion of the band. I prefer 144.074 MHz, but *anywhere between 144.000 to 144.100 is the spot for Moon bounce*. (WSJT EME can be found on 144.1-144.2 MHz.) **ANSWER B.**

E3A07 What frequency range would you normally tune to find EME signals in the 70 cm band?

A. 430.000 - 430.150 MHz.

B. 430.100 - 431.100 MHz.

C. 431.100 - 431.200 MHz.

D. 432.000 - 432.100 MHz.

Again, the 70-cm band is the 432-MHz band – *everything below 432.100* is within the weak-signal (only) portion of the band where CW and SSB can be found. **ANSWER D.**

E3A02 What characterizes libration fading of an Earth-Moon-Earth signal?

A. A slow change in the pitch of the CW signal.

B. A fluttery irregular fading.

C. A gradual loss of signal as the Sun rises.

D. The returning echo is several Hertz lower in frequency than the transmitted signal.

Remember, EME means Earth-Moon-Earth and that the Moon is not a perfect reflector of radio signals. All those craters on the Moon will cause your *return signal to sound fluttery, with rapid, irregular fading*. **ANSWER B.**

E3A03 When scheduling EME contacts, which of these conditions will generally result in the least path loss?

A. When the Moon is at perigee.

B. When the Moon is full.

C. When the Moon is at apogee.

D. When the MUF is above 30 MHz.

PERigee is the time when the Moon is closest to the Earth. This is the PERfect condition for EME to occur, and a PERfect day to schedule an EME contact. A *Moon at perigee*, just appearing above the horizon, will give you the *best opportunity for Moon-bounce*. **ANSWER A.**

Ground-mounted antennas, like this 10GHz EME dish in Alaska, should be surrounded by a wood safety fence. And make sure to stay a safe distance away from any antenna on transmit!

E3A05 Which of the following describes a method of establishing EME contacts?
A. Time synchronous transmissions with each station alternating.
B. Storing and forwarding digital messages.
C. Judging optimum transmission times by monitoring beacons from the Moon.
D. High speed CW identification to avoid fading.

The Moon is an elusive target to receive echoes. You don't just aim your huge antenna at the Moon and begin picking up signals! It just won't happen. You need to *synchronize your digital transmissions* to the Moon and *synchronize the reception* of another station sending back to you, with both of your chronometers set to WWV time, accurate down to the second. **ANSWER A.**

E2D03 Which of the following digital modes is especially useful for EME communications?
A. FSK441.   C. Olivia.
B. PACTOR III.   D. JT65.

It takes a mighty big long-boom antenna to bounce signals off the Moon, called Earth-Moon-Earth contacts. With voice, it is a huge effort, but going *digital with JT65* you can begin receiving Moon echoes with just a modest antenna system. You can hear digital sounds with echoes from the Moon on my exclusive audio CD course. Take a listen! **ANSWER D.**

E8C13 What is one advantage of using JT-65 coding?
A. Uses only a 65 Hz bandwidth.
B. The ability to decode signals which have a very low signal to noise ratio.
C. Easily copied by ear if necessary.
D. Permits fast-scan TV transmissions over narrow bandwidth.

Weak signal VHF/UHF and microwave operators know well the popular WSJT software to enable miraculous *reception of digital signals well below a normal noise floor. JT-65* coding allows your computer to capture everything from meteor scatter CQs to conversations off the surface of the Moon! Just because you can't hear anything coming out of the speaker, many times your computer's sound card program can deliver enough meaningful audio to decode programs that will utterly amaze you. **ANSWER B.**

Good to know:

Weak Signal Band Plan Recommendations

Over the past few years, tremendous advances have been made in weak-signal communications on the VHF and UHF bands using computer-generated data modes. These modes enable communications at or below the noise floor. Amateurs with modest stations are now making contacts over thousands of kilometers via EME and Meteor Scatter using these new modes. However, these very-weak data signals often cannot be detected by ear, which makes them incompatible with normal analog SSB or CW signals. There have been instances of interference caused by SSB operators who obviously can't hear the data signals. The recent EME Conference held in Germany recommended the creation of worldwide VHF and UHF sub-bands for JT-65, the weak signal data protocol. The suggested segments are 50.185 to 50.195 MHz on 6 meters; 144.115 to 144.135 MHz on 2 meters; 432.060 to 432.070 MHz on 70 centimeters; and 1296.060 to 1296.070 MHz.

E2D12 How does JT65 improve EME communications?

A. It can decode signals many dB below the noise floor using FEC.
B. It controls the receiver to track Doppler shift.
C. It supplies signals to guide the antenna to track the Moon.
D. All of these choices are correct.

The best digital emission for Earth Moon Earth (EME) communications is *JT65*, a data mode that *can detect JT65 signals down to 30 dB lower than your VHF/UHF noise floor! JT65 uses forward error correction (FEC)*, well-structured formats, a one minute transmit period, and message frames repeated up to five times. You can let your computer and transceiver work all night banging signals off the Moon, and when you wake up the next morning, Voila! Here are the contacts that took place! **ANSWER A.**

HERE ARE SOME WEB ADDRESSES OF USEFUL SITES RELATED TO OUTER SPACE COMMS:

http://spaceweather.com/
http://prop.hfradio.org/
www.amsat.org/amsat-new/news/
http://hamwaves.com/propagation/
http://www.spacew.com/www/realtime.php
http://www.arrl.org/news/arrl-committee-seeks-microwave-band-plan-input
http://www.arrl.org/news/space-station-active-on-70-cm-packet
http://pulsar.princeton.edu/~joe/k1jt
http://www.vhfdx.de/wsjt/
http://web.wt.net/~w5un/primer.htm
http://www.qsl.net/w8wn/hscw/hscwopen.html

Visuals & Video Modes

E2B01 How many times per second is a new frame transmitted in a fast-scan (NTSC) television system?

A. 30.

B. 60.

C. 90.

D. 120.

Operating fast-scan television is fun on the 420-MHz and higher bands. *A new frame is created 30 times per second* by two complete fields for each frame. This gives you a picture just like home television. In fact, older, analog, cable-ready TVs can still directly receive local ATV signals! **ANSWER A.**

Amateur Television

E2B02 How many horizontal lines make up a fast-scan (NTSC) television frame?
A. 30.
B. 60.
C. 525.
D. 1080.

A total of *525 lines from two fields make up a complete frame.* **ANSWER C.**

E2B03 How is an interlaced scanning pattern generated in a fast-scan (NTSC) television system?
A. By scanning two fields simultaneously.
B. By scanning each field from bottom to top.
C. By scanning lines from left to right in one field and right to left in the next.
D. By scanning odd numbered lines in one field and even numbered ones in the next.

Each field is 262.5 lines, and *the frame is created by scanning even numbered lines in one field and odd numbered lines in the second field.* **ANSWER D.**

E2B04 What is blanking in a video signal?
A. Synchronization of the horizontal and vertical sync pulses.
B. Turning off the scanning beam while it is traveling from right to left or from bottom to top.
C. Turning off the scanning beam at the conclusion of a transmission.
D. Transmitting a black and white test pattern.

The scanning beam is turned off when it completes its travel from left to right while it is returning from right to left to start another line. The scanning beam is also blanked when the beam reaches the bottom and is returning to the top to begin the next field. **ANSWER B.**

E2B08 Which of the following is a common method of transmitting accompanying audio with amateur fast-scan television?
A. Frequency-modulated sub-carrier.
B. A separate VHF or UHF audio link.
C. Frequency modulation of the video carrier.
D. All of these choices are correct.

Besides the standard 4.5 MHz sound subcarrier used by most ATV and analog broadcast stations, *there are a variety of ways to pass audio while transmitting an amateur radio fast-scan "live action" video signal.* To add audio to a fast-scan signal, we could frequency modulate the video carrier itself, known as "on carrier" sound, which was used in the early days when hams added video modulators to FM voice transceivers. Audio can also be simultaneously sent using FM audio on a dedicated VHF or UHF audio link in another band. This link also can be used as an ATV calling, coordination and talkback channel from the receiving stations. On the 2 meter band, most ATV stations use 144.340 MHz or 146.430 MHz for this link. **ANSWER D.** ☞ **Visit: www.hamtv.com**

E2B07 What is the name of the signal component that carries color information in NTSC video?
A. Luminance.
B. Chroma.
C. Hue.
D. Spectral Intensity.

Chroma is the video component contained in a phase modulated 3.58 MHz sub-carrier that varies with the instantaneous color in the scene being scanned by the video camera. **ANSWER B.**

E2B06 What is vestigial sideband modulation?

A. Amplitude modulation in which one complete sideband and a portion of the other are transmitted.

B. A type of modulation in which one sideband is inverted.

C. Narrow-band FM transmission achieved by filtering one sideband from the audio before frequency modulating the carrier.

D. Spread spectrum modulation achieved by applying FM modulation following single sideband amplitude modulation.

Vestigal sideband modulation *(VSB) is amplitude modulation of a carrier in which one complete sideband is transmitted* (upper VSB is standard) *and a vestigal sideband filter is used to attenuate the other sideband* video energy greater than 1.25 MHz and the 3.58 MHZ color and 4.5 MHz sound sub-carrier frequencies. This helps reduce occupied bandwidth to 6 MHz and provides better spectrum efficiency than full double sideband modulation. **ANSWER A.**

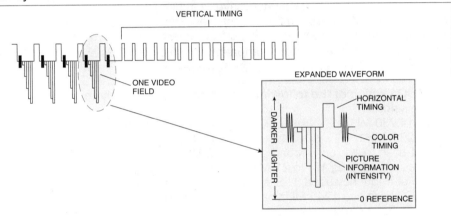

a. A TV Composite Video Signal

Source: *Using Video in Your Home,* G. McComb. © 1989, Master Publishing, Inc., Niles, IL.

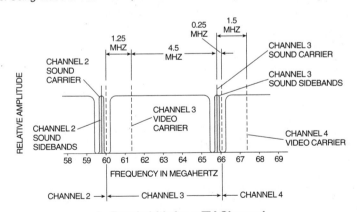

b. Bandwidth for a TV Channel

Source: *Antennas,* A. Evans, K. Britain,. © 1988, Master Publishing, Inc., Niles, IL.

The video signal is composed of three major components: picture, timing, and color. Television sets "know" how to decode this stream of information into pictures that reproduce the changing scenes captured by the video camera.

E2B05 Which of the following is an advantage of using vestigial sideband for standard fast-scan TV transmissions?
A. The vestigial sideband carries the audio information.
B. The vestigial sideband contains chroma information.
C. Vestigial sideband reduces bandwidth while allowing for simple video detector circuitry.
D. Vestigial sideband provides high frequency emphasis to sharpen the picture.

Vestigal sideband (VSB) cuts off the lower color and sound sub-carriers at -3.58 and -4.50 MHz from the video carrier, respectively, as they are unnecessary. Therefore the frequencies below 1.25 MHz from the video carrier in a VSB transmitter will have very low energy radiated and this frees up room for more users of the band. Total TV channel *bandwidth* is +4.75 / -1.25 MHz from the video carrier equaling 6 MHz. *Simple analog video detectors are used* in the TV receiver with most of the work being done by the VSB filter in the IF ahead of the detector. Even though commercial TV signals have switched from analog to digital, many ham fast-scan TV signals remain analog. New TVs will still employ an analog tuner and vestigal sideband receiver with a simple video detector in addition to the digital system, for some time, to allow compatibility with older VCRs and DVD players. **ANSWER C.** ☞ **Visit: www.hampubs.com/atv.php**

E2B18 On which of the following frequencies is one likely to find FM ATV transmissions?
A. 14.230 MHz.
B. 29.6 MHz.
C. 52.525 MHz.
D. 1255 MHz.

FM fast-scan television is found on the 902 MHz and higher bands because it occupies a whopping 17- to 21-MHz of bandwidth, depending on the sound subcarrier frequency used. U.S. standard is 4 MHz deviation and 5.5 MHz sound in the 902 MHz and 1240 MHz bands, resulting in an occupied bandwidth of 19 MHz. The center frequency must then be at least 9.5 MHz from the band edge so the sideband energy does not fall outside the band limits. *1255 MHz* is a reasonable frequency to use given these parameters. The beauty of FM ATV, thanks to signal-to-noise limiter multiplication, is a snow-free picture that can be received up to 12 dB weaker in signal strength than an AM ATV picture. **ANSWER D.**
☞ **Visit: www.ac6rb.com**

E2B16 Which of the following is the video standard used by North American fast-scan ATV stations?
A. PAL.
B. DRM.
C. Scottie.
D. NTSC.

North American fast-scan analog ATV stations transmit *NTSC, National Television System Committee*, identical to the old analog, over-the-air broadcast TV signals. Broadcast is now all digital. PAL is the analog broadcast TV system used primarily in Europe and other countries. DRM and Scottie are SSTV modes. The question asks for the North American fast-scan video standard, which is NTSC. **ANSWER D.**

E2B12 How are analog SSTV images typically transmitted on the HF bands?
A. Video is converted to equivalent Baudot representation.
B. Video is converted to equivalent ASCII representation.
C. Varying tone frequencies representing the video are transmitted using PSK.
D. Varying tone frequencies representing the video are transmitted using single sideband.

Transmitting a slow scan picture requires not much more than your HF transceiver, a sound card, ultra–inexpensive SSTV software, and your favorite camcorder to freeze frame your smiling mug. The *varying tone frequencies* you hear on a transmitted signal, on SSB, *represent the video* going out over the airwaves. To receive a color SSTV signal, all you need to add to your HF transceiver is a patch cable to your computer's sound card mic input, and the rest is accomplished mostly by the SSTV software! **ANSWER D.**

E2B17 What is the approximate bandwidth of a slow-scan TV signal?

A. 600 Hz.   C. 2 MHz.
B. 3 kHz.   D. 6 MHz.

We can find slow scan television on both high frequency as well as on VHF/UHF. On high frequency, look for *slow scan* at 14.230 MHz, and notice that it takes only *3 kHz of bandwidth*, but up to 45 seconds for a complete single picture. **ANSWER B.**

E2B19 What special operating frequency restrictions are imposed on slow scan TV transmissions?

A. None; they are allowed on all amateur frequencies.
B. They are restricted to 7.245 MHz, 14.245 MHz, 21.345, MHz, and 28.945 MHz.
C. They are restricted to phone band segments and their bandwidth can be no greater than that of a voice signal of the same modulation type.
D. They are not permitted above 54 MHz.

Slow-scan TV transmissions, as well as facsimile, are allowed in phone band segments as long as their *bandwidth is no greater than that of a voice signal* of the same modulation type. You can easily hear 20-meter slow-scan signals by tuning in 14.230 and 14.233, a hot-bed of SSTV activity that is easily seen on a computer program or on your little handheld SSTV transmit/receive equipment. **ANSWER C.**

Amateur slow-scan TV pictures can be displayed
on a tube, or an LCD display like this one

E2B13 How many lines are commonly used in each frame on an amateur slow-scan color television picture?

A. 30 to 60.   C. 128 or 256.
B. 60 or 100.   D. 180 or 360.

It takes *128 lines to make up a complete frame. 256 lines increases the resolution* of the color. **ANSWER C.**

E2B15 What signals SSTV receiving equipment to begin a new picture line?
A. Specific tone frequencies.
B. Elapsed time.
C. Specific tone amplitudes.
D. A two-tone signal.

Depending on which SSTV mode you are using, *new picture lines are triggered by specific tone frequencies.* **ANSWER A.**

E2B14 What aspect of an amateur slow-scan television signal encodes the brightness of the picture?
A. Tone frequency.
B. Tone amplitude.
C. Sync amplitude.
D. Sync frequency.

Brightness of an SSTV picture depends on the *higher tone frequencies* received. **ANSWER A.** ☞ **Visit: www.bonito.net**

E2B09 What hardware, other than a receiver with SSB capability and a suitable computer, is needed to decode SSTV using Digital Radio Mondiale (DRM)?
A. A special IF converter.
B. A special front end limiter.
C. A special notch filter to remove synchronization pulses.
D. No other hardware is needed.

The term "Digital Radio Mondiale" *(DRM) is not a special kind of circuit* with a big chip in the middle. Rather, DRM is a panel of commercial and non-commercial radio users who have developed the sole standard for sending broadcasts over worldwide low-frequency and high-frequency bands. SSTV is easy to hear at 14.230 MHz, and it's easy to capture a full-color, high resolution picture with just your HF transceiver and a suitably-equipped computer with a sound card running SSTV software. It's easy to get great pictures on high frequency. **ANSWER D.**

E2B10 Which of the following is an acceptable bandwidth for Digital Radio Mondiale (DRM) based voice or SSTV digital transmissions made on the HF amateur bands?
A. 3 KHz.   C. 15 KHz.
B. 10 KHz.   D. 20 KHz.

The typical high-frequency, *slow-scan signal* occupies no more room than a conventional single sideband voice signal, *3 kHz or less*. **ANSWER A.**

E2B11 What is the function of the Vertical Interval Signaling (VIS) code transmitted as part of an SSTV transmission?
A. To lock the color burst oscillator in color SSTV images.
B. To identify the SSTV mode being used.
C. To provide vertical synchronization.
D. To identify the call sign of the station transmitting.

There are over a dozen individual methods of sending a color slow scan television signal over HF and VHF airwaves. The most common are Martin and Scottie. These popular modes require the initial reception of *the 300 ms VIS code, which is necessary for your computer to know which SSTV mode it is being sent*. **ANSWER B.**

E6D04 **What function does a charge-coupled device (CCD) serve in a modern video camera?**

A. It stores photogenerated charges as signals corresponding to pixels.
B. It generates the horizontal pulses needed for electron beam scanning.
C. It focuses the light used to produce a pattern of electrical charges corresponding to the image.
D. It combines audio and video information to produce a composite RF signal.

The charge-coupled device is continually clocking the data out, or the data "dissipates" like a dynamic RAM. *The CCD converts photo-generated charges into signals corresponding to pixels.* **ANSWER A.**

E6D03 **Which of the following is true of a charge-coupled device (CCD)?**

A. Its phase shift changes rapidly with frequency.
B. It is a CMOS analog-to-digital converter.
C. It samples an analog signal and passes it in stages from the input to the output.
D. It is used in a battery charger circuit.

The charge-coupled device *(CCD) samples an analog signal and pass it in stages from input to output.* **ANSWER C.**

Amateur TV signals can be received on a variety
of equipment – even a small hand-held monitor

E6D14 **Which is NOT true of a charge-coupled device (CCD)?**

A. It uses a combination of analog and digital circuitry.
B. It can be used to make an audio delay line.
C. It is commonly used as an analog-to-digital converter.
D. It samples and stores analog signals.

This question asks which is NOT true of a *CCD*. A charge-coupled device *cannot be used as an analog-to-digital converter.* **ANSWER C.**

☞ Visit: http://electronics.howstuffworks.com/camcorder2.htm

E6D01 What is cathode ray tube (CRT) persistence?

A. The time it takes for an image to appear after the electron beam is turned on.
B. The relative brightness of the display under varying conditions of ambient light.
C. The ability of the display to remain in focus under varying conditions.
D. The length of time the image remains on the screen after the beam is turned off.

The term "*persistence*" in a cathode-ray tube *is the length of time that an image will remain on the screen* after the beam is turned off. **ANSWER D.**

E6D13 What type of CRT deflection is better when high-frequency waveforms are to be displayed on the screen?

A. Electromagnetic.   C. Radar.
B. Tubular.   D. Electrostatic.

Cathode-ray tube *(CRT) electrostatic deflection* is better for high-frequency signals. Horizontal and vertical deflection plates, similar to capacitor plates, deflect the electron beam onto the screen according to signal voltages applied to the plates. **ANSWER D.**

E6D02 Exceeding what design rating can cause a cathode ray tube (CRT) to generate X-rays?

A. The heater voltage.   C. The operating temperature.
B. The anode voltage.   D. The operating frequency.

X-rays can be emitted from a cathode-ray tube *if the anode voltage is increased* beyond the value at which the tube is safely rated. **ANSWER B.**

E6D05 What is a liquid-crystal display (LCD)?

A. A modern replacement for a quartz crystal oscillator which displays its fundamental frequency.
B. A display that uses a crystalline liquid which, in conjunction with polarizing filters, becomes opaque when voltage is applied.
C. A frequency-determining unit for a transmitter or receiver.
D. A display that uses a glowing liquid to remain brightly lit in dim light.

A *liquid-crystal display*, found on most amateur radio handheld units, *uses a crystalline liquid in conjunction with polarizing filters* to change the way light is refracted from a rear mirror through the liquid. **ANSWER B.** ☞ **Visit: www.hamtv.org**

Liquid Crystal Displays are popular because of their large size and low power consumption

E6D15 What is the principle advantage of liquid-crystal display (LCD) devices over other types of display devices?

A. They consume less power.
B. They can display changes instantly.
C. They are visible in all light conditions.
D. They can be easily interchanged with other display devices.

It takes *very little current* to sustain a liquid crystal display segment. **ANSWER A.**

Digital Excitement with Your Computers & Radios

E2E01 Which type of modulation is common for data emissions below 30 MHz?
A. DTMF tones modulating an FM signal.
B. FSK.
C. Pulse modulation.
D. Spread spectrum.

Frequency shift keying (FSK) is created by data pulses from your computer that shift the master oscillator back and forth to create two distinct tones in the other operator's receiver. Audio Frequency Shift Keying (AFSK) generates audio tones from your computer's sound card or multi-mode controller and is what we use for more complex digital signaling like PSK 31, MFSK, and CLOVER. These modes have more than two distinct frequency signals. Both *Frequency Shift Keying (FSK)* and Audio Frequency Shift Keying (AFSK) can be found on HF, *below 30 MHz* . Listen to my audio CDs to hear the difference between these two modes.
ANSWER B.

E2E11 What is the difference between direct FSK and audio FSK?
A. Direct FSK applies the data signal to the transmitter VFO.
B. Audio FSK has a superior frequency response.
C. Direct FSK uses a DC-coupled data connection.
D. Audio FSK can be performed anywhere in the transmit chain.

With *direct frequency shift keying* (FSK) the *variable frequency oscillator* in your transceiver *changes the actual frequency* in sync with the 1s and 0s data from your computer. With audio frequency shift keying (AFSK) your master oscillator stays on frequency, and audio tones are generated by a sound card or modem to correspond precisely with the data being sent. **ANSWER A.**

E2E07 What is the typical bandwidth of a properly modulated MFSK16 signal?
A. 31 Hz.   C. 550 Hz.
B. 316 Hz.   D. 2.16 kHz.

MFSK refers to a super-charged RTTY signal, Millennium Frequency Shift Keying. Instead of simple RTTY with two tones, MFSK 16 uses 16 tones, one at a time at 15.625 baud, spaced at 15.625 Hertz apart. This works out to a relatively narrow *316 Hertz bandwidth*, just like RTTY, and offers higher non-error rates with improved reception under noisy conditions. It is an ideal mode for 40-, 80-, and 160-meters. **ANSWER B.**

E2E06 What is the most common data rate used for HF packet communications?
A. 48 baud.   C. 300 baud.
B. 110 baud.   D. 1200 baud.

High frequency packet must labor along at a relatively slow *300 baud* so these signals do not exceed bandwidth limits below 30 MHz. 300 baud seems relatively slow compared to higher speeds on VHF and UHF, and high-frequency packet is not necessarily a "right now" transmission method. However, it is highly accurate, and even binary files may be sent around the world via HF packet! **ANSWER C.**

E2D09 Under clear communications conditions, which of these digital communications modes has the fastest data throughput?
A. AMTOR.   C. PSK31.
B. 170-Hz shift, 45 baud RTTY. D. 300-baud packet.

Although *300-baud packet* sounds relatively slow, on the worldwide bands it's faster than most other modes. **ANSWER D.**

E2D02 What is the definition of baud?
A. The number of data symbols transmitted per second.
B. The number of characters transmitted per second.
C. The number of characters transmitted per minute.
D. The number of words transmitted per minute.

The term *"baud" is defined as the number of data symbols transmitted per second*. With Baudot signaling speeds, the character count will not match the data symbol rate. We call those symbols transmitted per second as a number of events or discreet conditions per second. In the Baudot code, one character may be represented by five bits. **ANSWER A.**

E2E08 Which of the following HF digital modes can be used to transfer binary files?

A. Hellschreiber.   C. RTTY.
B. PACTOR.   D. AMTOR.

PACTOR is more than 20 years old, and the transmission sounds like 2 birds chirping. Data is sent in blocks and the receiving station sends ACK for acknowledging everything sent, or NAK indicating the request to re-send the block. *PACTOR may be used to transfer binary files*, allowing you to download programs from a BBS and then work the program on your laptop or P.C. **ANSWER B.**

E8C02 What are some of the differences between the Baudot digital code and ASCII?

A. Baudot uses four data bits per character, ASCII uses seven or eight; Baudot uses one character as a shift code, ASCII has no shift code.
B. Baudot uses five data bits per character, ASCII uses seven or eight; Baudot uses two characters as shift codes, ASCII has no shift code.
C. Baudot uses six data bits per character, ASCII uses seven or eight; Baudot has no shift code, ASCII uses two characters as shift codes.
D. Baudot uses seven data bits per character, ASCII uses eight; Baudot has no shift code, ASCII uses two characters as shift codes.

Baudot is a 5-bit code; ASCII is a 7-bit code. In actuality, ASCII is an 8-bit code with only 7 bits uniformly defined. There are *2 shift codes in Baudot*, an up-shift and a down-shift. **ANSWER B.**

E2E02 What do the letters FEC mean as they relate to digital operation?

A. Forward Error Correction.   C. Fatal Error Correction.
B. First Error Correction.   D. Final Error Correction.

The word "AMTOR" stands for amateur teleprinting over radio. This is a digital mode with error correction leading to perfect copy when band conditions are favorable on worldwide frequencies. *"FEC" stands for forward error correction*, AMTOR Mode B, where each character is sent twice. **ANSWER A.**

E2E03 How is Forward Error Correction implemented?

A. By the receiving station repeating each block of three data characters.
B. By transmitting a special algorithm to the receiving station along with the data characters.
C. By transmitting extra data that may be used to detect and correct transmission errors.
D. By varying the frequency shift of the transmitted signal according to a predefined algorithm.

During AMTOR FEC, *each data character is sent twice* with no acknowledgment by the receiving station. **ANSWER C.**

E2E05 How does ARQ accomplish error correction?

A. Special binary codes provide automatic correction.
B. Special polynomial codes provide automatic correction.
C. If errors are detected, redundant data is substituted.
D. If errors are detected, a retransmission is requested.

ARQ, (Automatic Repeat reQuest) is a digital protocol that acknowledges individual data blocks and *requests a repeat of any missed block*. ARQ is part of the PACTOR and CLOVER digital signaling systems. **ANSWER D.**

E2E04 What is indicated when one of the ellipses in an FSK crossed-ellipse display suddenly disappears?

A. Selective fading has occurred.
B. One of the signal filters has saturated.
C. The receiver has drifted 5 kHz from the desired receive frequency.
D. The mark and space signal have been inverted.

Most high-frequency stations feature a companion station monitor oscilloscope that will display two crossed ellipses, indicating a properly tuned-in digital signal. *If one of the ellipses disappears*, the phenomenon known as *selective fading has occurred.* This will cause your received copy to stall until the signal comes out of the fade and is then printed out on the screen or on paper. Don't touch the receiver during selective fading – you are on frequency; it's just that propagation is causing the signal to fade in and out. **ANSWER A.**

E8C03 What is one advantage of using the ASCII code for data communications?

A. It includes built-in error-correction features.
B. It contains fewer information bits per character than any other code.
C. It is possible to transmit both upper and lower case text.
D. It uses one character as a shift code to send numeric and special characters.

ASCII code allows upper and lower case text to be transmitted; Baudot does not. **ANSWER C.**

Bit Position											
				0	1	0	1	1	0	0	1
				0	0	1	1	1	1	0	0
1				1	1	1	1	0	0	0	0
0	0	0	0	@	P	‘	p	0	sp	NUL	DLE
1	0	0	0	A	Q	a	q	1	!	SOH	DC1
0	1	0	0	B	R	b	r	2	"	STX	DC2
1	1	0	0	C	S	c	s	3	#	ETX	DC3
0	0	1	0	D	T	d	t	4	$	EOT	DC4
1	0	1	0	E	U	e	u	5	%	ENQ	NAK
0	1	1	0	F	Z	f	v	6	&	ACK	SYN
1	1	1	0	G	W	g	w	7	’	BEL	ETB
0	0	0	1	H	X	h	x	8	(	BS	CAN
1	0	0	1	I	Y	i	y	9	)	HT	EM
0	1	0	1	J	Z	j	z	:	*	LF	SUB
1	1	0	1	K	[	k	{	;	+	VT	ESC
0	0	1	1	L	\	l	|	<	,	FF	FS
1	0	1	1	M	]	m	}	=	-	CR	GS
0	1	1	1	N	^	n	~	>	.	SO	RS
1	1	1	1	O	_	o	DEL	?	/	SI	US

American Standard Code for Information Interchange (ASCII)

E8C12 What is the advantage of including a parity bit with an ASCII character stream?

A. Faster transmission rate.
B. The signal can overpower interfering signals.
C. Foreign language characters can be sent.
D. Some types of errors can be detected.

The parity bit is used for error detection when sending data. A parity bit 1 or 0 might indicate the odd or even number of 1s or 0s in the data block. It is not foolproof – if a pair of 1s or a pair of 0s are missed, the error detection circuit might consider no errors. Yet this simple error detection method is commonly used and works surprisingly well to confirm that the other station received the character intact, or needs a repeat. **ANSWER D.**

E8C06 What is the necessary bandwidth of a 170-hertz shift, 300-baud ASCII transmission?

A. 0.1 Hz.
B. 0.3 kHz.
C. 0.5 kHz.
D. 1.0 kHz.

Digital bandwidth is:

BW = baud rate + (1.2 × f shift)
BW = 300 + (1.2 × 170) = 504 Hz = *0.504 kHz*. **ANSWER C.**

E8C07 What is the necessary bandwidth of a 4800-Hz frequency shift, 9600-baud ASCII FM transmission?

A. 15.36 kHz.
B. 9.6 kHz.
C. 4.8 kHz.
D. 5.76 kHz.

Digital bandwidth again – use the formula at question E8C06. In this problem, the baud rate is 9600 and the frequency shift is 4800. *Bandwidth = 9600 + (1.2 × 4800) = 15360*. The answer is 15,360 Hz, and when the decimal point is moved three places to the left, the bandwidth works out to be *15.36 kHz*. **ANSWER A.**

E8C04 What technique is used to minimize the bandwidth requirements of a PSK31 signal?

A. Zero-sum character encoding.
B. Reed-Solomon character encoding.
C. Use of sinusoidal data pulses.
D. Use of trapezoidal data pulses.

Phase shift keying at 31 bit rate – actually 31.25 – uses sinusoidal data pulses, allows for synchronous reception through the demodulation of the signal with no delay, to the signal delayed by 1 bit as polarity is shifted. The number "0" represents a phase reversal and a "1" represents the steady carrier. The PSK-31 VARICODE allocates just a few 1s and 0s for the more common letters, and a long string of 1s and 0s for less used characters. For example, E is Varicode 11 while Q is Varicode 110111111. Some new, big ticket HF transceivers now include a *PSK-31 decoder* for on-screen, no-computer-required reading of messages. The *use of 31 Hz sinusoidal data pulses* provides sidebands with carrier phase excursions of less than 31.25 Hz, resulting in near-perfect copy using very little bandwidth. One critical component is a rock-solid receiver that can follow the signal with no less than 2 or 3 Hz error. **ANSWER C.** ☞ **Visit: www.ykc.com/wa5ufh/wsjtgroup**

E2E09 Which of the following HF digital modes uses variable-length coding for bandwidth efficiency?

A. RTTY.   C. MT63.
B. PACTOR.   D. PSK31.

PSK31 stands for Phase Shift Keying with a bit rate slightly over 31. It is a little bit like Morse code, with the more common letters represented by a short count of bits for that character:

| E 11 | A 1011 |
| T 101 | O 111 |

The less used characters, such as Q and X, might contain up to 10 bits. This *Varicode is transmitted* with audio phase shifts, occupying a mere 31 Hz of bandwidth, leading to bandwidth efficiency. When *using PSK31*, you cannot copy PSK 63 signals, although they might be visible on the waterfall as wider PSK31 signals. Similarly, when using PSK63, you might see more narrow PSK 31 signals on the waterfall, but you will not be able to copy them without switching to PSK31. PSK63 is just like PSK31, except that it is twice as fast and twice as wide. Macros and Brag files will transmit at 100 wpm on PSK 63, instead of 50 wpm on PSK31. Using PSK63, the turnover and exchange times for contesting will be twice as fast as on PSK31, and even faster than RTTY. Weak signal performance of PSK 63 is similar to PSK31, except that twice the power needs to be run compared to PSK31 for the same signal-to-noise ratio. **ANSWER D.**

E2E10 Which of these digital communications modes has the narrowest bandwidth?

A. MFSK16.   C. PSK31.
B. 170-Hz shift, 45 baud RTTY.   D. 300-baud packet.

Phase shift keying is the type of modulation that generates PSK, and the "31" means the bit rate is 31.25. There is no mistaking the whistle sound of *PSK 31*, and the modern transceiver with adaptive digital signal processing can pull a signal out of the noise when you think you might not get any reception at all. If you have a modern ham set and a computer, working PSK 31 at around 50 wpm requires only a relatively inexpensive matching device and the readily-available software and a sound card. You will be fascinated when you eavesdrop on two operators running PSK 31, and I expect that it will ultimately become one of the most popular computer-to-computer digital modes, going well beyond RTTY. **ANSWER C.**

E4A09 Which of the following describes a good method for measuring the intermodulation distortion of your own PSK signal?

A. Transmit into a dummy load, receive the signal on a second receiver, and feed the audio into the sound card of a computer running an appropriate PSK program.
B. Multiply the ALC level on the transmitter during a normal transmission by the average power output.
C. Use an RF voltmeter coupled to the transmitter output using appropriate isolation to prevent damage to the meter.
D. All of these choices are correct.

When sending digital signals, any stage within your transmitter that is not absolutely linear may generate intermodulation distortion, giving other stations trying to receive your PSK 31 signal a tough time. If you don't have an oscilloscope, you could *transmit a test message on low power into a dummy load* while receiving the signal on a second receiver, usually with no antenna connected to avoid overload. That second receiver then outputs to a sound card with your computer running the PSK program, and there are several options that let you see your transmitted signal to insure your IMD (InterModulation Distortion) is at its lowest point. **ANSWER A.**

E4A11 Which of these instruments could be used for detailed analysis of digital signals?
    A. Dip meter.
    B. Oscilloscope.
    C. Ohmmeter.
    D. Q meter.

You need to exercise precise level adjustments to send out digital signals. Too much gain will cause distortion. Too little gain may cause your signal to be missed by the receiving station. The use of an *oscilloscope* lets you confirm both linearity of the transmitted data signals as well as signal levels. Also watch your transceiver's ALC meter to insure it is barely moving. **ANSWER B.**

E2D10 How can an APRS station be used to help support a public service communications activity?
    A. An APRS station with an emergency medical technician can automatically transmit medical data to the nearest hospital.
    B. APRS stations with General Personnel Scanners can automatically relay the participant numbers and time as they pass the check points.
    C. An APRS station with a GPS unit can automatically transmit information to show a mobile station's position during the event.
    D. All of these choices are correct.

Many marathons are supported by ham operators on foot, bicycles, and motorcycles with an *APRS station tied into a GPS to automatically transmit position information* along the course route. All this can be done without the operator needing to do a thing – simply set the time interval you want your APRS packets transmitted and, presto!, everyone with the right equipment can easily see the APRS position information. **ANSWER C.** ☞ **Visit: www.findu.com**

Kenwood dual bander plugged into the Avmap G5 GPS position plotter.

E2D07 Which of the following digital protocols is used by APRS?
A. PACTOR. C. AX.25.
B. 802.11. D. AMTOR.

The digital packet protocol in use for *APRS is* described as AX.25, the ham version of commercial packet networks operating X.25. The *AX.25 protocol* is recognized and authorized by the Federal Communications Commission with specific speed limitations depending on what band you are operating on. **ANSWER C.**

E2D08 Which of the following types of packet frames is used to transmit APRS beacon data?
A. Unnumbered Information. C. Acknowledgement.
B. Disconnect. D. Connect.

The *packet frames* used to transmit APRS beacon data *are called UIF –
Unnumbered Information Frames*. Networking systems can allow an APRS position and short message to be transmitted and relayed throughout the country. One manufacturer produces a dual-band handheld with the APRS terminal node controller already built-in. Handheld GPS equipment that outputs NMEA data is available for under $99, and mapping GPS can also show ham call signs and locations. **ANSWER A.** ☞ **Visit: www.geostat.us**

E2D11 Which of the following data are used by the APRS network to communicate your location?
A. Polar coordinates. C. Radio direction finding LOPs.
B. Time and frequency. D. Latitude and longitude.

It is *latitude and longitude* in degrees, minutes, and fractions of a minute, that *are transmitted by your APRS* radio system, usually on 144.390 MHz on the 2 meter band. The data transmitted is usually in the NMEA 0183 format for most APRS systems, and I-Gate stations tied into the internet may take your GPS position and have it displayed at http://www.aprs.fi. **ANSWER D.** ☞ **Visit: www.aprs.fi**

E2D06 Which of the following is a commonly used 2-meter APRS frequency?
A. 144.390 MHz. C. 145.020 MHz.
B. 144.200 MHz. D. 146.520 MHz.

APRS stands for Automatic Position Reporting System. APRS links a portable or fixed-mount Global Positioning System (GPS) receiver to a ham transmitter or transceiver. APRS allows the ham radio to transmit pre-set timed packet bursts of the GPS position, either on worldwide high frequencies, or on the VHF and UHF bands. *On 2 meters, the most common APRS frequency is 144.390 MHz*. **ANSWER A.**

E8C08 What term describes a wide-bandwidth communications system in which the transmitted carrier frequency varies according to some predetermined sequence?
A. Amplitude compandored single sideband.
B. AMTOR.
C. Time-domain frequency modulation.
D. Spread-spectrum communication.

Spread spectrum communications are becoming more popular. You may not realize it, but your new 900-MHz cordless phone uses spread-spectrum communications. The frequency hops in a pre-arranged sequence, eliminating interference and eavesdropping. **ANSWER D.**

E1F01 On what frequencies are spread spectrum transmissions permitted?
  A. Only on amateur frequencies above 50 MHz.
  B. Only on amateur frequencies above 222 MHz.
  C. Only on amateur frequencies above 420 MHz.
  D. Only on amateur frequencies above 144 MHz.

Spread spectrum is still new to ham radio operators, and *is confined to the 222 MHz band and above.* For the time being, spread spectrum is not allowed on 2- or 6-meters. [97.305] **ANSWER B.**

900MHz cordless phones use spread-spectrum technology.

E8C09 Which of these techniques causes a digital signal to appear as wide-band noise to a conventional receiver?
  A. Spread-spectrum.
  B. Independent sideband.
  C. Regenerative detection.
  D. Exponential addition.

Do you own a GPS receiver? If you could listen to the 1575 MHz, the GPS public channel, all you would hear is noise. It takes a specific *spread spectrum* receiver to translate this *wide-band noise* into meaningful reception. Ham radio operators are allowed spread spectrum from 222 MHZ and above. Power output of 100 watts is allowed, but any more than 1 watt of spread spectrum requires automatic transmitter power control circuits that identify a strong signal and reduce its own transmit power. Spread spectrum (SS) is just one more exciting "new mode" that hams are encouraged to explore above 222 MHz. **ANSWER A.**

E8C10 What spread-spectrum communications technique alters the center frequency of a conventional carrier many times per second in accordance with a pseudo-random list of channels?
  A. Frequency hopping.
  B. Direct sequence.
  C. Time-domain frequency modulation.
  D. Frequency compandored spread-spectrum.

One spread spectrum communications technique alters the center frequency of an RF carrier in a pseudo-random manner causing the frequency to "hop" around from channel to channel. This is termed *"frequency hopping."* **ANSWER A.**

E2C09 How does the spread-spectrum technique of frequency hopping work?

    A. If interference is detected by the receiver it will signal the transmitter to change frequencies.

    B. If interference is detected by the receiver it will signal the transmitter to wait until the frequency is clear.

    C. A pseudo-random binary bit stream is used to shift the phase of an RF carrier very rapidly in a particular sequence.

    D. The frequency of the transmitted signal is changed very rapidly according to a particular sequence also used by the receiving station.

Frequency hopping depends on a particular signaling sequence, and the frequency of the RF carrier is changed so quickly that it is almost indistinguishable except to other spread-spectrum receivers locked on to the same signaling sequence. **ANSWER D.**

E8C11 What spread-spectrum communications technique uses a high speed binary bit stream to shift the phase of an RF carrier?

    A. Frequency hopping.

    B. Direct sequence.

    C. Binary phase-shift keying.

    D. Phase compandored spread-spectrum.

The term "*direct sequence*" is used to describe a system where a *fast binary stream shifts* the phase of an RF carrier. This is another spread spectrum communications technique. Using only a single carrier frequency and then modulating the signal by shifting the phase in accordance with a high-speed binary bit stream results in a signal whose pattern on a spectrum analyzer spreads out into a pattern of hills and valleys. This is called direct sequence spread spectrum. **ANSWER B.**

E2C08 Why are received spread-spectrum signals resistant to interference?

    A. Signals not using the spectrum-spreading algorithm are suppressed in the receiver.

    B. The high power used by a spread-spectrum transmitter keeps its signal from being easily overpowered.

    C. The receiver is always equipped with a digital blanker circuit.

    D. If interference is detected by the receiver it will signal the transmitter to change frequencies.

Amateur operators transmitting and receiving spread-spectrum signals are just pioneering one more way that hams can serve our valuable frequency resources. Although spread-spectrum may utilize an entire band, the hopping technique won't disturb other signals on the band, and all other *signals not using the same spread-spectrum algorithm are suppressed in the receiver*. So just when you thought there was no more room for any more signals on a particular band, spread-spectrum gets through! **ANSWER A.**

E1F10 What is the maximum transmitter power for an amateur station transmitting spread spectrum communications?

    A. 1 W.

    B. 1.5 W.

    C. 10 W.

    D. 1.5 kW.

Recent FCC rule changes now limit 222 MHz and up *spread spectrum power output to no more than 10 watts*. Spread spectrum occupies a large amount of bandwidth, but it has its own ability to dodge FM and digital signals that may be within the spread spectrum frequency range. Cell phones are a good example of spread spectrum. Spread spectrum may use code-division multiple access, direct sequence, or frequency hopping. **ANSWER C.**

E1F09 Which of the following conditions apply when transmitting spread spectrum emission?

    A. A station transmitting SS emission must not cause harmful interference to other stations employing other authorized emissions.
    B. The transmitting station must be in an area regulated by the FCC or in a country that permits SS emissions.
    C. The transmission must not be used to obscure the meaning of any communication.
    D. All of these choices are correct.

Spread-spectrum communications are *allowed to any other station under FCC regulation, and* to *stations in foreign countries* that have specifically permitted spread-spectrum communications to their country from ours. Spread-spectrum *may not be used to obscure the meaning of any communication*. If these 3 requirements are met, and as long as you stay at no more than 10 watts, you are entitled to use spread spectrum on 222 MHz and up. [97.311] **ANSWER D.**

E2E12 Which type of digital communication does not support keyboard-to-keyboard operation?

    A. Winlink.
    B. RTTY.
    C. PSK31.
    D. MFSK.

It's fun to exchange messages with other hams using your keyboard in real time. You can hear the sounds of RTTY (Radio TeleTYpe), PSK 31 (Phase Shift Keying) and MFSK (Multi Frequency Shift Keying) on my lively audio CD course. This is keyboard-to-keyboard at its best! Winlink is a great way to send digital traffic into the e-mail system, using PACTOR protocols. Winlink systems can also send graphics and weather picture charts. But *Winlink is not designed for keyboard-to-keyboard communications*. **ANSWER A.**

E8C05 What is the necessary bandwidth of a 13-WPM international Morse code transmission?

    A. Approximately 13 Hz.   C. Approximately 52 Hz.
    B. Approximately 26 Hz.   D. Approximately 104 Hz.

To determine the bandwidth of a Morse code signal, use this formula – bandwidth equals:

$$BW_{cw} = baud\ rate \times wpm \times fading\ factor$$

For CW, the baud rate is 0.8 and the fading factor is 5, therefore, *multiply 0.8 times 13 wpm times 5. The answer works out to be 52 Hz*. **ANSWER C.**

E8C01 Which one of the following digital codes consists of elements having unequal length?

A. ASCII.

C. Baudot.

B. AX.25.

D. Morse code.

Morse Code is made up of unequal length elements. Dits are short, and dahs are long. **ANSWER D.** ☞ **Visit: www.dxatlas.com/morserunner**

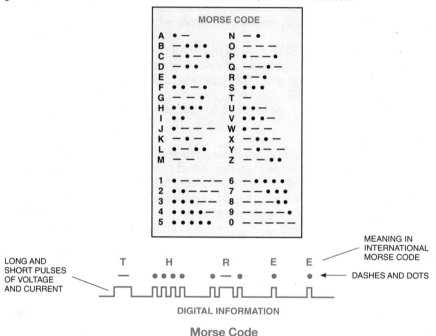

Morse Code

HERE ARE SOME WEB ADDRESSES OF USEFUL SITES RELATED TO DIGITAL EXCITEMENT YOU'RE YOUR COMPUTERS & RADIOS:

www.dstarinfo.com/

www.dstarinfo.com/repeater-lists.aspx

www.dstarusers.org/repeaters.php

www.cq-amateur-radio.com

www.dvdongle.com

www.rsgbiota.org/

www.dxpub.com/dx_mag.html

www.dailydx.com

http://dxcluster.ham-radio.ch

www.dstarinfo.com

www.dstarinfo.com/repeater-lists.aspx

www.dstarusers.org/repeaters.php

www.cq-amateur-radio.com/cq-awards/cq-waz-wards/index_cq_waz_award.html

---

Modulate Your Transmitters

E8B01 What is the term for the ratio between the frequency deviation of an RF carrier wave and the modulating frequency of its corresponding FM-phone signal?

    A. FM compressibility.
    B. Quieting index.
    C. Percentage of modulation.
    D. Modulation index.

Be careful with this one. Since they don't say the highest modulating frequency, but say only "modulating frequency," they are looking for the *modulation index*. Notice that there is no answer that says "deviation ratio" to get you in trouble. **ANSWER D.**

E8B02 How does the modulation index of a phase-modulated emission vary with RF carrier frequency (the modulated frequency)?

    A. It increases as the RF carrier frequency increases.
    B. It decreases as the RF carrier frequency increases.
    C. It varies with the square root of the RF carrier frequency.
    D. It does not depend on the RF carrier frequency.

This is one of those answers that has a "not" in it that makes it correct. When you calculate the *modulation index*, you *do not look at the actual RF carrier frequency* as part of your calculations. **ANSWER D.**

FM versus AM

Greater Number of Sidebands

In FM modulation, one sine wave is used to modify another sine wave and multiple sums and differences result, depending on the deviation ratio. This is illustrated in Figure a. Sidebands appear on either side of f_c at f_m, $2f_m$, $3f_m$, etc., depending upon the deviation ratio. Figure b shows that there are five significant sidebands for a deviation ratio of 3, and Figure c shows three significant sidebands for a deviation ratio of 1.67. The multiple sidebands are formed in both the upper and lower sideband positions, separated by the modulating frequency.

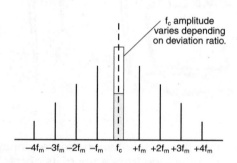

a. General Spectrum

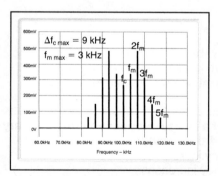

b. Deviation Ratio of 3

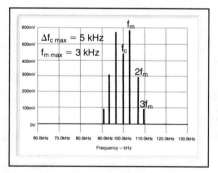

c. Deviation Ratio of 1.67

Frequency spectrum of FM signals.

Source: *Basic Communications Electronics*, Hudson & Luecke, © 1999 Master Publishing, Inc., Niles, IL

E8B03 What is the modulation index of an FM-phone signal having a maximum frequency deviation of 3000 Hz either side of the carrier frequency, when the modulating frequency is 1000 Hz?

    A. 3.   C. 3000.

    B. 0.3.   D. 1000.

3000 divided by 1000 gives you a *modulation index of 3*. **ANSWER A.**

$$\text{Modulation Index } (\chi) = \frac{\text{Peak Deviation (D)}}{\text{Modulation Frequency } (m)} \text{ or } \chi = \frac{D}{m}$$

E8B04 What is the modulation index of an FM-phone signal having a maximum carrier deviation of plus or minus 6 kHz when modulated with a 2-kHz modulating frequency?

    A. 6000.   C. 2000.

    B. 3.   D. 1/3.

6 divided by 2 gives us a *modulation index of 3*. **ANSWER B.**

E1B12 What is the highest modulation index permitted at the highest modulation frequency for angle modulation?

    A. 0.5.   C. 2.0.

    B. 1.0.   D. 3.0.

With angle modulation, a sine wave carrier is varied by various angles to replicate the incoming modulation source. The *highest modulation index permitted for angle modulation is exactly 1.0*. **ANSWER B.**

E8B09 What is meant by deviation ratio?

    A. The ratio of the audio modulating frequency to the center carrier frequency.

    B. The ratio of the maximum carrier frequency deviation to the highest audio modulating frequency.

    C. The ratio of the carrier center frequency to the audio modulating frequency.

    D. The ratio of the highest audio modulating frequency to the average audio modulating frequency.

It's important to set the deviation ratio properly on your FM transceiver to prevent splatter and a signal bandwidth that is too wide. *Deviation ratio* is the ratio of the *maximum carrier swing to your highest audio modulating frequency* when you speak into the microphone. **ANSWER B.**

E8B05 What is the deviation ratio of an FM-phone signal having a maximum frequency swing of plus-or-minus 5 kHz when the maximum modulation frequency is 3 kHz?

    A. 60.   C. 0.6.

    B. 0.167.   D. 1.67.

Deviation ratio is the ratio of the maximum carrier swing to your highest audio modulating frequency when you speak into the microphone. To calculate the deviation ratio, divide the maximum carrier frequency swing by the maximum modulation frequency. *3 kHz modulation frequency into 5 kHz frequency swing gives you a deviation ratio of 1.67*. **ANSWER D.**

$$\text{Deviation Ratio} = \frac{\text{Maximum Carrier Frequency Deviation in kHz}}{\text{Maximum Modulation Frequency in kHz}}$$

E8B06 What is the deviation ratio of an FM-phone signal having a maximum frequency swing of plus or minus 7.5 kHz when the maximum modulation frequency is 3.5 kHz?

    A. 2.14.
    B. 0.214.
    C. 0.47.
    D. 47.

To calculate deviation ratio of an FM signal, simply divide the maximum carrier frequency swing by the maximum modulation frequency. The small number goes into the larger number. *3.5 kHz into 7.5 kHz ends up with a deviation ratio of 2.14.* Simple, huh? **ANSWER A.**

E8B08 What parameter does the modulating signal vary in a pulse-position modulation system?

    A. The number of pulses per second.
    B. The amplitude of the pulses.
    C. The duration of the pulses.
    D. The time at which each pulse occurs.

This question is about a pulse-position modulation system. The position is the time at which each pulse occurs. The *modulating signal varies the time when each pulse occurs.* **ANSWER D.**

E8B07 When using a pulse-width modulation system, why is the transmitter's peak power greater than its average power?

    A. The signal duty cycle is less than 100%.
    B. The signal reaches peak amplitude only when voice modulated.
    C. The signal reaches peak amplitude only when voltage spikes are generated within the modulator.
    D. The signal reaches peak amplitude only when the pulses are also amplitude modulated.

In pulse-width modulation, the output signal is composed of pulses of carrier wave of varying widths. Each pulse is at the transmitter's peak power, but between pulses the transmitter is at zero power. Since pulse modulation is not a continuous burst of power, the *duty cycle is always less than 100 percent.* **ANSWER A.**

E8B10 Which of these methods can be used to combine several separate analog information streams into a single analog radio frequency signal?

    A. Frequency shift keying.
    B. A diversity combiner.
    C. Frequency division multiplexing.
    D. Pulse compression.

In *frequency division multiplexing (FDM),* a single carrier is modulated by multiple signals simultaneously. While this is a viable digital technique in the telecommunications industry, its use on ham bands at VHF, UHF, and microwave frequencies usually is for point-to-point retransmissions of multiple conversations to conserve precious bandwidth. It is likely that converted commercial equipment would be required for this type of point-to-point signal forwarding. **ANSWER C.**

E8B11 Which of the following describes frequency division multiplexing?

    A. The transmitted signal jumps from band to band at a predetermined rate.

    B. Two or more information streams are merged into a "baseband," which then modulates the transmitter.

    C. The transmitted signal is divided into packets of information.

    D. Two or more information streams are merged into a digital combiner, which then pulse position modulates the transmitter.

The *"information streams" are the baseband*. Each information stream is modulated onto a sub-carrier (up-converting it), which is then used to modulate the carrier. In the receiver, the sub-carriers are removed by pass band filters, then each information stream is recovered by down-converting the sub-carrier to baseband. **ANSWER B.**

E8B12 What is digital time division multiplexing?

    A. Two or more data streams are assigned to discrete sub-carriers on an FM transmitter.

    B. Two or more signals are arranged to share discrete time slots of a data transmission.

    C. Two or more data streams share the same channel by transmitting time of transmission as the sub-carrier.

    D. Two or more signals are quadrature modulated to increase bandwidth efficiency.

Time division multiplexing (TDM) allows multiple users to share the same transmit and receive channel, divided into *discreet time slots*, for the sending of data. Again, this modern form of signaling is most used by commercial and telecommunications point-to-point broadcasting, but hams with the right equipment could take advantage of this frequency conserving technology. The two signals don't actually share discrete time slots – rather, they are separated by each using a different time slot. **ANSWER B.**

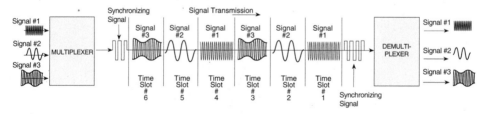

Time multiplexing and demultiplexing.

Source: *Basic Communications Electronics*, Hudson & Luecke, © 1999 Master Publishing, Inc., Niles, IL

E8D01 Which of the following is the easiest voltage amplitude parameter to measure when viewing a pure sine wave signal on an analog oscilloscope?

    A. Peak-to-peak voltage.

    B. RMS voltage.

    C. Average voltage.

    D. DC voltage.

Looking at a pure sine wave signal on a scope, it is easy to identify the peak-to-peak voltage measured from the maximum positive excursion of the signal to the maximum negative excursion of the signal. **ANSWER A.**

Peak, Peak-to-Peak and RMS Voltages

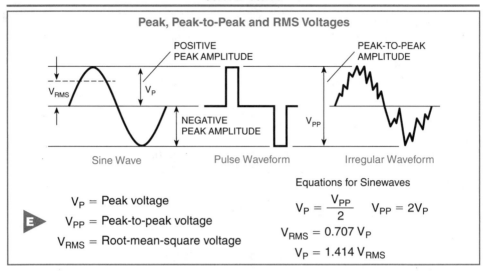

POSITIVE PEAK AMPLITUDE

PEAK-TO-PEAK AMPLITUDE

V_{RMS}

V_P

NEGATIVE PEAK AMPLITUDE

V_{PP}

Sine Wave   Pulse Waveform   Irregular Waveform

Equations for Sinewaves

V_P = Peak voltage

V_{PP} = Peak-to-peak voltage

V_{RMS} = Root-mean-square voltage

$$V_P = \frac{V_{PP}}{2} \qquad V_{PP} = 2V_P$$

$$V_{RMS} = 0.707\, V_P$$

$$V_P = 1.414\, V_{RMS}$$

E8D02 **What is the relationship between the peak-to-peak voltage and the peak voltage amplitude of a symmetrical waveform?**

A. 0.707:1.   C. 1.414:1.

B. 2:1.   D. 4:1.

The relationship between peak-to-peak voltage and peak voltage is 2:1. It is important to realize that the waveform must be symmetrical – and a sine wave certainly is. $V_{PP} = 2V_P$. **ANSWER B.**

E8A01 **What type of wave is made up of a sine wave plus all of its odd harmonics?**

A. A square wave.   C. A cosine wave.

B. A sine wave.   D. A tangent wave.

If you're *odd*, you're probably *square*. **ANSWER A.**

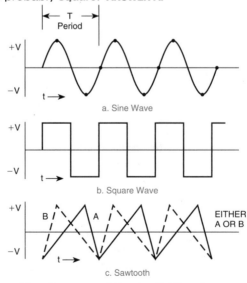

T
Period

+V

−V

t →

a. Sine Wave

+V

−V

t →

b. Square Wave

+V

B   A

EITHER A OR B

−V

t →

c. Sawtooth

Sine, Square and Sawtooth Waveforms

E8A02 What type of wave has a rise time significantly faster than its fall time (or vice versa)?

A. A cosine wave.
B. A square wave.
C. A sawtooth wave.
D. A sine wave.

The *sawtooth wave* can be backwards, too – an immediate rise time significantly faster than the fall time. **ANSWER C.**

E8A03 What type of wave is made up of sine waves of a given fundamental frequency plus all its harmonics?

A. A sawtooth wave.
B. A square wave.
C. A sine wave.
D. A cosine wave.

If it has *all the harmonics* the answer is a *sawtooth* wave. **ANSWER A.**

E8A04 What is equivalent to the root-mean-square value of an AC voltage?

A. The AC voltage found by taking the square of the average value of the peak AC voltage.
B. The DC voltage causing the same amount of heating in a given resistor as the corresponding peak AC voltage.
C. The DC voltage causing the same amount of heating in a resistor as the corresponding RMS AC voltage.
D. The AC voltage found by taking the square root of the average AC value.

You sometimes see this as RMS, also called effective value of an AC voltage. Since AC voltage is continuously changing in amplitude, we must look at its *root mean square value* before we can come up with the *same heating effect in a given resistor as a constant DC voltage*. **ANSWER C.**

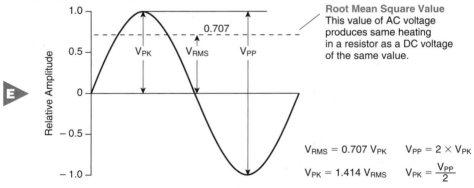

Root Mean Square Value
This value of AC voltage produces same heating in a resistor as a DC voltage of the same value.

$V_{RMS} = 0.707\ V_{PK}$ $\qquad V_{PP} = 2 \times V_{PK}$

$V_{PK} = 1.414\ V_{RMS}$ $\qquad V_{PK} = \dfrac{V_{PP}}{2}$

RMS (V_{RMS}), Peak (V_{PK}), and Peak-to-Peak (V_{PP}) Voltage

E8A05 What would be the most accurate way of measuring the RMS voltage of a complex waveform?

A. By using a grid dip meter.
B. By measuring the voltage with a D'Arsonval meter.
C. By using an absorption wavemeter.
D. By measuring the heating effect in a known resistor.

In a very complex wave form, one way to determine its effective value is to *measure the heating effect in a known resistor*, compared to a known DC voltage. **ANSWER D.**

E8A07 What determines the PEP-to-average power ratio of a single-sideband phone signal?

A. The frequency of the modulating signal.
B. The characteristics of the modulating signal.
C. The degree of carrier suppression.
D. The amplifier gain.

It's not all that easy to determine the ratio of peak envelope power to average power in your worldwide SSB set because your speech characteristics (the *characteristics of the modulating signal*) may vary from those of another operator when using the same equipment. **ANSWER B.**

On SSB transmit, peak envelope power (PEP) is determined by your speech characteristics

E8A06 What is the approximate ratio of PEP-to-average power in a typical single-sideband phone signal?

A. 2.5 to 1.
B. 25 to 1.
C. 1 to 1.
D. 100 to 1.

When you talk about power output from your SSB worldwide set, we normally refer to power as peak envelope power. Depending on your modulation, *the ratio of PEP to average power is about 2.5 to 1.* You can remember 2.5 because that's very close to 2.5 kHz of bandwidth occupied by a properly-modulated SSB signal. **ANSWER A.**

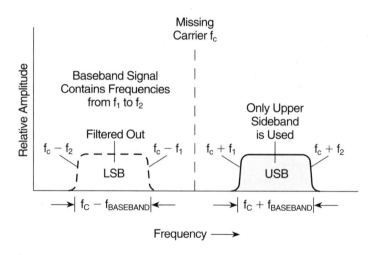

Spectral plot of an SSB signal.

Source: *Basic Communications Electronics*, Hudson & Luecke, © 1999 Master Publishing, Inc., Niles, IL

E8D10 What type of meter should be used to monitor the output signal of a voice-modulated single-sideband transmitter to ensure you do not exceed the maximum allowable power?

A. An SWR meter reading in the forward direction.
B. A modulation meter.
C. An average reading wattmeter.
D. A peak-reading wattmeter.

The *peak-reading watt meter* momentarily holds the highest amount of output signal in order for you to double check that you are not exceeding the maximum allowable power. Many newer, worldwide ham sets have a peak-reading power output meter with an LCD display built in. **ANSWER D.**

E8D06 What is the advantage of using a peak-reading wattmeter to monitor the output of a SSB phone transmitter?

A. It is easier to determine the correct tuning of the output circuit.
B. It gives a more accurate display of the PEP output when modulation is present.
C. It makes it easier to detect high SWR on the feed line.
D. It can determine if any flat-topping is present during modulation peaks.

When you transmit on your brand-new, high-frequency transceiver you got yourself as a reward for passing your Extra Class exam, it is likely you will see that transceiver's average power output indicator not going much above 70 watts as you are speaking. There is nothing wrong with you equipment – you may be transmitting 100 watts peak envelope power with the meter showing only about 70 watts due to the latency in the meter movement, or the slower reaction time that some LCD meters employ to average out their readings. A *peak-reading watt meter* compensates for this seemingly lower-than-usual power output indication and faithfully gathers enough power output information on your varying voice waveform to show you your *true peak envelope power (PEP) output*. Most transceivers have an aggressive ALC (automatic limiting control) modulation setting, so don't expect any panel meter built in to the gear to instantly fly off the scale above 100 watts. **ANSWER B.**

E8D04 What is the PEP output of a transmitter that develops a peak voltage of 30 volts into a 50-ohm load?

A. 4.5 watts.
B. 9 watts.
C. 16 watts.
D. 18 watts.

The first step in solving this problem is to take the peak voltage and reduce it to average voltage by multiplying $30 \times 0.707 = 21.21$ volts. Next, use the following formula for average power:

Average volts $\times$ average current = average power
Average power, $P = E^2/R$
$P = (21.21)^2 \div 50 = 8.99$ watts

ANSWER B.

E8D11 What is the average power dissipated by a 50-ohm resistive load during one complete RF cycle having a peak voltage of 35 volts?

A. 12.2 watts.
B. 9.9 watts.
C. 24.5 watts.
D. 16 watts.

The first step in solving this problem is to take the peak voltage and reduce it to the RMS voltage by multiplying 35 volts $\times$ 0.707 = 24.745 volts.

Average volts $\times$ average current = average power
Average power, $P = E^2/R$
P = (24.745)² ÷ 50 = 12.2463 watts, or 12.2 watts

ANSWER A.

E8D12 What is the peak voltage of a sinusoidal waveform if an RMS-reading voltmeter reads 34 volts?

A. 123 volts.
B. 96 volts.
C. 55 volts.
D. 48 volts.

In this question, *34 volts RMS is multiplied by 1.414 = 48.076 peak volts.* Remember, to go from average to peak, multiple by 1.414. Going from peak to average, you would multiply by 0.707, as we did in the earlier problem. **ANSWER D.**

E8D13 Which of the following is a typical value for the peak voltage at a standard U.S. household electrical outlet?

A. 240 volts.
B. 170 volts.
C. 120 volts.
D. 340 volts.

When you measure your house line voltage with a volt meter, you are measuring *RMS*. If you were to look at it on an oscilloscope, and look at the electrical peaks, it *would be 1.414 times higher than 120 volts, or a total of 170 volts peak*. **ANSWER B.**

E8D14 Which of the following is a typical value for the peak-to-peak voltage at a standard U.S. household electrical outlet?

A. 240 volts.
B. 120 volts.
C. 340 volts.
D. 170 volts.

If you really want to impress your friends about your house line voltage, tell them that you have *340 volts* coming out of the socket. If they bet you don't, tell them you are stating the voltage as peak-to-peak, which is twice the peak voltage. **ANSWER C.**

E8D15 Which of the following is a typical value for the RMS voltage at a standard U.S. household electrical power outlet?

A. 120V AC.
B. 340V AC.
C. 85V AC.
D. 170V AC.

Household AC line voltage is rated at 120-V AC RMS. **ANSWER A.**

E8D16 What is the RMS value of a 340-volt peak-to-peak pure sine wave?

A. 120V AC.
B. 170V AC.
C. 240V AC.
D. 300V AC.

340 volts peak-to-peak divided in half is 170 volts peak. Multiply the peak voltage by 0.707, and you end up with *120 volts AC, RMS.* **ANSWER A.**

E8D05 If an RMS-reading AC voltmeter reads 65 volts on a sinusoidal waveform, what is the peak-to-peak voltage?
    A. 46 volts.   C. 130 volts.
    B. 92 volts.   D. 184 volts.
$Vp = 1.414 \times V_{RMS}$, and $Vpp = 2Vp$. We will first change 65 volts RMS to peak voltage by multiplying $65 \times 1.414 = 91.91$. Since they want peak-to-peak voltage, *multiply this by 2*, and you end up with *184 volts*. **ANSWER D.**

E8A08 What is the period of a wave?
    A. The time required to complete one cycle.
    B. The number of degrees in one cycle.
    C. The number of zero crossings in one cycle.
    D. The amplitude of the wave.
The *period of a wave* is the *time required to complete one cycle of that wave*. See the illustration at E8A01. It is the reciprocal of the frequency $T = 1 \div f$. If the frequency is 1 MHz, then $T = 1 \div 1MHz = 1\mu s$. If a wave goes through one million cycles in one second, then each cycle takes 1 one-millionth of a second. **ANSWER A.**

E8A09 What type of waveform is produced by human speech?
    A. Sinusoidal.   C. Irregular.
    B. Logarithmic.   D. Trapezoidal.
Sound waves between 300 and 3000 Hz are the most useful by radio modulator circuits to transmit the human voice. Unlike pure tones going into the microphone circuit, our *human speech* word tones form *irregular waves* with constantly changing amplitude and frequency. **ANSWER C.**

E8A13 What is an advantage of using digital signals instead of analog signals to convey the same information?
    A. Less complex circuitry is required for digital signal generation and detection.
    B. Digital signals always occupy a narrower bandwidth.
    C. Digital signals can be regenerated multiple times without error.
    D. All of these choices are correct.
An advantage of digital signals is the capability of *relaying the signals multiple times without error*. Digital signals in a noisy environment may also be forward-error-corrected, giving your equipment and computer a second chance at decoding an incoming noisy data signal. Many times you can't even hear the data signal, yet your equipment can faithfully decode that incoming digital signal. Noise isn't absent in digital signals, but the fact that the signal is either in one state or another (sometimes either on or off, or other times at one discrete frequency or another) means that it is easier to detect the exact state it is in. In comparison, analog signals vary over a very wide range of amplitudes and frequencies, and thus are more prone to becoming lost among the noise and interference. Because when you recover a digital signal you recover only the discrete states, you can faithfully reproduce the original signal without distortion. **ANSWER C.**

E8A12 What type of information can be conveyed using digital waveforms?
    A. Human speech.   C. Data.
    B. Video signals.   D. All of these choices are correct.
Even analog voice may be converted into data, just like what happens inside your cell phone. Over-the-air television signals are now digital and, of course, data is digital, so *all of the answers are correct*. **ANSWER D.**

E8A14 Which of these methods is commonly used to convert analog signals to digital signals?
- A. Sequential sampling.
- B. Harmonic regeneration.
- C. Level shifting.
- D. Phase reversal.

Changing an analog signal into a digital signal takes place in the analog-to-digital converter through *sequential sampling* of the waveform. Thousands of tiny audio slices are held in a sequential sample-and-hold circuit where they are converted into binary numbers. Some ADCs count with a staircase generator while others convert analog voltage into digital values with multiple comparators, yet others in DSS use a phase accumulator to identify points on a specific waveform cycle. Sequential sampling of the incoming waveform is the heart of converting analog signals to digital. **ANSWER A.**

E8A15 What would the waveform of a stream of digital data bits look like on a conventional oscilloscope?
- A. A series of sine waves with evenly spaced gaps.
- B. A series of pulses with varying patterns.
- C. A running display of alpha-numeric characters.
- D. None of the above; this type of signal cannot be seen on a conventional oscilloscope.

When you look at a *digital signal on a conventional oscilloscope*, you will see a series of *sharp pulses with varying patterns*, illustrating a string of data bits, 1s and 0s. **ANSWER B.**

E8A10 Which of the following is a distinguishing characteristic of a pulse waveform?
- A. Regular sinusoidal oscillations.
- B. Narrow bursts of energy separated by periods of no signal.
- C. A series of tones that vary between two frequencies.
- D. A signal that contains three or more discrete tones.

Sending Morse code is a good example of a pulse waveform — *short bursts of energy separated by spaces*. Intermittent Continuous Wave (ICW) still continues to be a popular amateur radio activity, even though the FCC has completely eliminated any Morse code testing. **ANSWER B.**

E8A11 What is one use for a pulse modulated signal?

A. Linear amplification.	C. Multiphase power transmission.
B. PSK31 data transmission.	D. Digital data transmission.

Pulse modulation is used for some exciting data transmissions:

CW	WINLINK 2000
RTTY	HELLSCHREIBER
PACTOR	G-TOR
MFSK (MultiFrequency Shift Keying)	CLOVER

Many of the new high frequency transceivers may already have DSP capabilities to decode some of these *digital data transmissions*. Add a laptop or home computer to your radio and enjoy all of the excitement on your new Extra class sub-bands transmitting data. Remember, keep your power levels reduced, and keep your data gain controls at a minimum. **ANSWER D.**

Modulate Your Transmitters

E7E01 Which of the following can be used to generate FM phone emissions?
- A. A balanced modulator on the audio amplifier.
- B. A reactance modulator on the oscillator.
- C. A reactance modulator on the final amplifier.
- D. A balanced modulator on the oscillator.

The *reactance modulator* causes the *oscillator* to vary frequency in accordance with the modulation. It's the oscillator that is *modulated*. **ANSWER B.**

E7E02 What is the function of a reactance modulator?
- A. To produce PM signals by using an electrically variable resistance.
- B. To produce AM signals by using an electrically variable inductance or capacitance.
- C. To produce AM signals by using an electrically variable resistance.
- D. To produce PM signals by using an electrically variable inductance or capacitance.

The *reactance modulator* is found in a frequency modulation transceiver, *varying inductance or capacitance to produce* the *phase modulation (PM)* signal where the phase of the carrier current is varied with the frequency of the modulating voltage. **ANSWER D.**

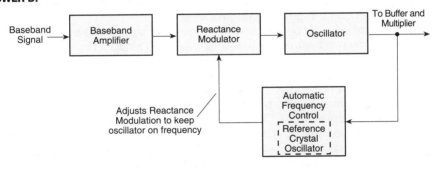

Block diagram for generating FM modulation

Source: *Basic Communications Electronics*, Hudson & Luecke, © 1999 Master Publishing, Inc., Niles, IL

E7E03 How does an analog phase modulator function?
- A. By varying the tuning of a microphone preamplifier to produce PM signals.
- B. By varying the tuning of an amplifier tank circuit to produce AM signals.
- C. By varying the tuning of an amplifier tank circuit to produce PM signals.
- D. By varying the tuning of a microphone preamplifier to produce AM signals.

Phase modulation is developed by *varying the tuning of the amplifier tank circuit to produce phase modulation*, but never in the microphone preamp circuit. **ANSWER C.**

E7E05 What circuit is added to an FM transmitter to boost the higher audio frequencies?
- A. A de-emphasis network.
- B. A heterodyne suppressor.
- C. An audio prescaler.
- D. A pre-emphasis network.

Pre-emphasis networks in modern FM transmitters help improve the signal-to-noise ratio by proportionally attenuating the less-important lower audio frequencies for voice, and *boosting* voice intelligence in the *higher audio frequencies*. **ANSWER D.**

E7E04 What is one way a single-sideband phone signal can be generated?
A. By using a balanced modulator followed by a filter.
B. By using a reactance modulator followed by a mixer.
C. By using a loop modulator followed by a mixer.
D. By driving a product detector with a DSB signal.

In a *single-sideband transceiver*, a *lower or upper sideband filter* removes the unwanted sideband. **ANSWER A.**

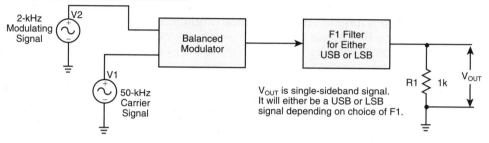

Generating an SSB signal

E7E13 Which of the following describes a common means of generating an SSB signal when using digital signal processing?
A. Mixing products are converted to voltages and subtracted by adder circuits.
B. A frequency synthesizer removes the unwanted sidebands.
C. Emulation of quartz crystal filter characteristics.
D. The quadrature method.

Digital signal processing has revolutionized ham radio transmitters and receivers, and a host of after market audio products, to help clean up signal reception. For SSB signal transmissions, *DSP relies on phase-shifting* carrier and audio components by 90 degrees, and sending this phased split to the SSB balanced modulator. The output is a DSP upper or lower sideband, with the carrier and unwanted sideband suppressed more than 40 dB. No heterodyning is required, and all of this *quadrature occurs in the DSP chip*. **ANSWER D.**

E7C09 What type of digital signal processing filter might be used to generate an SSB signal?
A. An adaptive filter.   C. A Hilbert-transform filter.
B. A notch filter.   D. An elliptical filter.

In a *DSP* unit for transmit, we might use the *Hilbert-transform filter* – think of "transform" *for SSB transmit*. **ANSWER C.**

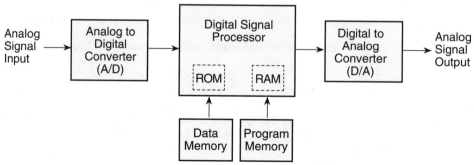

Block diagram of a basic digital signal processing (DSP) system

E7E07 What is meant by the term baseband in radio communications?
- A. The lowest frequency band that the transmitter or receiver covers.
- B. The frequency components present in the modulating signal.
- C. The unmodulated bandwidth of the transmitted signal.
- D. The basic oscillator frequency in an FM transmitter that is multiplied to increase the deviation and carrier frequency.

The term *"baseband" refers to the range of frequencies occupied by the modulating signal* in the transmitter. For digital modes and CW, it could be as narrow as 30 Hz, and for sideband voice and FM, baseband frequency response could range from 2.5 kHz up to 25 kHz. **ANSWER B.**

E1B01 Which of the following constitutes a spurious emission?
- A. An amateur station transmission made at random without the proper call sign identification.
- B. A signal transmitted to prevent its detection by any station other than the intended recipient.
- C. Any transmitted bogus signal that interferes with another licensed radio station.
- D. An emission outside its necessary bandwidth that can be reduced or eliminated without affecting the information transmitted.

May your spurious emissions be minimum! The FCC says so. A *spurious emission* could include random splatter to adjacent frequencies, second harmonics, and any other microvolt or millivolt emission that serves no purpose *outside of your necessary bandwidth*. [97.3] **ANSWER D.**

E1B08 What limitations may the FCC place on an amateur station if its signal causes interference to domestic broadcast reception, assuming that the receiver(s) involved are of good engineering design?
- A. The amateur station must cease operation.
- B. The amateur station must cease operation on all frequencies below 30 MHz.
- C. The amateur station must cease operation on all frequencies above 30 MHz.
- D. The amateur station must avoid transmitting during certain hours on frequencies that cause the interference.

If the FCC gets involved in an interference issue with your neighbor and it finds that your neighbor has a nice clean receiver of modern design, and your 1.5 kilowatt output is coming in over that receiver, *the FCC could impose quiet hours on your transmissions on the specific band* where you are causing the interference. Suggest you don't run the KW during ball games! [97.121(a)] **ANSWER D.**

E7B12 What type of circuit is shown in Figure E7-1?
- A. Switching voltage regulator.
- B. Linear voltage regulator.
- C. Common emitter amplifier.
- D. Emitter follower amplifier.

In a *common emitter amplifier*, the input signal is applied between base and emitter. In this circuit, the emitter is at AC ground because of C3. The output is taken from the collector. **ANSWER C.**

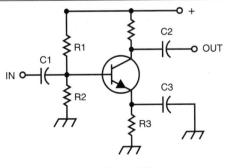

Figure E7-1

E7B10 In Figure E7-1, what is the purpose of R1 and R2?
A. Load resistors.
B. Fixed bias.
C. Self bias.
D. Feedback.

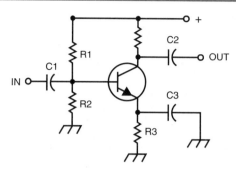

Figure E7-1

This reminds you a little bit of Thevenin's Theorem doesn't it? *R1 and R2* form a voltage divider to set the voltage and the current at the base of the transistor for *fixed bias.* **ANSWER B.**

E7B11 In Figure E7-1, what is the purpose of R3?
A. Fixed bias.
B. Emitter bypass.
C. Output load resistor.
D. Self bias.

Current through *R3 produces* a voltage drop that affects the base-emitter bias voltage. Since the transistor's collector and base current (the emitter current) determine the voltage at the emitter, it is called *self bias.* **ANSWER D.**

E7B13 In Figure E7-2, what is the purpose of R?
A. Emitter load.
B. Fixed bias.
C. Collector load.
D. Voltage regulation.

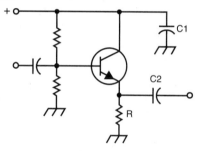

Figure E7-2

One end of the resistor R is connected to the emitter, the other end to ground. It is classified as an *emitter load resistor* in this circuit because the output is taken from the emitter. **ANSWER A.**

E7B14 In Figure E7-2, what is the purpose of C2?
A. Output coupling.
B. Emitter bypass.
C. Input coupling.
D. Hum filtering.

Notice the little circle after C2? This denotes the output of the circuit, and *C2 is the output coupling capacitor.* **ANSWER A.**

E7C01 How are the capacitors and inductors of a low-pass filter Pi-network arranged between the network's input and output?
A. Two inductors are in series between the input and output, and a capacitor is connected between the two inductors and ground.
B. Two capacitors are in series between the input and output and an inductor is connected between the two capacitors and ground.
C. An inductor is connected between the input and ground, another inductor is connected between the output and ground, and a capacitor is connected between the input and output.
D. A capacitor is connected between the input and ground, another capacitor is connected between the output and ground, and an inductor is connected between input and output.

The *pi-network features a capacitor in parallel with the input*, a second *capacitor in parallel with the output*, and an *inductor in series between the two*. It looks like the Greek letter π. **ANSWER D.**

E7C02 A T-network with series capacitors and a parallel shunt inductor has which of the following properties?
    A. It is a low-pass filter.
    B. It is a band-pass filter
    C. It is a high-pass filter.
    D. It is a notch filter.

The *T-network* transforms impedances and is also used as a *high-pass filter*. **ANSWER C.**

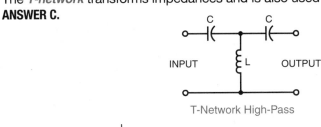

T-Network High-Pass

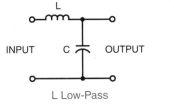

L Low-Pass

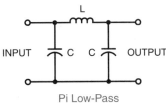

Pi Low-Pass

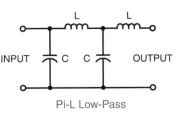

Pi-L Low-Pass

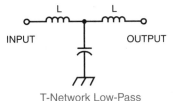

T-Network Low-Pass

Matching Networks

E7C03 What advantage does a Pi-L-network have over a Pi-network for impedance matching between the final amplifier of a vacuum-tube transmitter and an antenna?
    A. Greater harmonic suppression.
    B. Higher efficiency.
    C. Lower losses.
    D. Greater transformation range.

Worldwide ham sets transmitting into a multi-band antenna system can generate harmonics. The *pi-L-network* always *gives us greater harmonic suppression* inside the transceiver. **ANSWER A.**

E7C11 Which of the following is the common name for a filter network which is equivalent to two L networks connected back-to-back with the inductors in series and the capacitors in shunt at the input and output?
    A. Pi-L.
    B. Cascode.
    C. Omega.
    D. Pi.

The *Pi network* looks like the Greek letter π, and consists of two variable capacitors, shunt to ground, for transmitter output impedance matching. The Pi network also has only a single inductor. **ANSWER D.**

E7C12 Which of the following describes a Pi-L network used for matching a vacuum-tube final amplifier to a 50-ohm unbalanced output?
A. A Phase Inverter Load network.
B. A Pi network with an additional series inductor on the output.
C. A network with only three discrete parts.
D. A matching network in which all components are isolated from ground.

The pi-L network is characterized by two series-adjustable inductors, and two variable shunt-to-ground capacitors. You can spot the pi-L network over a simple pi network as the *pi-L network has an additional series inductor at the output*. **ANSWER B.**

E7C13 What is one advantage of a Pi matching network over an L matching network consisting of a single inductor and a single capacitor?
A. The Q of Pi networks can be varied depending on the component values chosen.
B. L networks cannot perform impedance transformation.
C. Pi networks have fewer components.
D. Pi networks are designed for balanced input and output.

The big advantage of a *Pi matching network* over the L network is the shunt-to-ground variable capacitors that *can vary the Q*, depending on the value of the variable capacitors and adjustable tap coil chosen. **ANSWER A.**

E7B09 Which of the following describes how the loading and tuning capacitors are to be adjusted when tuning a vacuum tube RF power amplifier that employs a pi-network output circuit?
A. The loading capacitor is set to maximum capacitance and the tuning capacitor is adjusted for minimum allowable plate current.
B. The tuning capacitor is set to maximum capacitance and the loading capacitor is adjusted for minimum permissible plate current.
C. The loading capacitor is adjusted to minimum plate current while alternately adjusting the tuning capacitor for maximum allowable plate current.
D. The tuning capacitor is adjusted for minimum plate current, while the loading capacitor is adjusted for maximum permissible plate current.

If you're going to be running a big power amplifier on the HF bands, chances are it will still use a vacuum tube for kilowatts of power output. Solid-state power amplifiers require no *tuning*, but most solid-state power amps only put out about 800 watts max. With a *tube power amplifier*, alternately tune and *increase the plate current with a loading capacitor*, and then *dip the plate current with a tuning capacitor*. Do this for short periods of time, and do it on a frequency that no one else is transmitting on. I first tune up into a dummy load to get me close in settings, and then switch over to the antenna, select a frequency which no one is using, announce, "I'm going to quickly tune up," and then quickly tune up. **ANSWER D.**

E7C04 How does an impedance-matching circuit transform a complex impedance to a resistive impedance?
A. It introduces negative resistance to cancel the resistive part of impedance.
B. It introduces transconductance to cancel the reactive part of impedance.
C. It cancels the reactive part of the impedance and changes the resistive part to a desired value.
D. Network resistances are substituted for load resistances and reactances are matched to the resistances.

In older worldwide sets, you could vary the matching network to *cancel the reactive (X) part of an impedance and change the value of the resistive part (R) of an impedance*. Newer transistorized sets have a fixed output, and there is no manual operator control of the fixed matching devices. However, manufacturers have come to the rescue in antenna matching by providing new worldwide ham sets with built-in automatic antenna tuners. **ANSWER C.**

E4D03 How can intermodulation interference between two repeaters occur?
    A. When the repeaters are in close proximity and the signals cause feedback in the final amplifier of one or both transmitters.
    B. When the repeaters are in close proximity and the signals mix in the final amplifier of one or both transmitters.
    C. When the signals from the transmitters are reflected out of phase from airplanes passing overhead.
    D. When the signals from the transmitters are reflected in phase from airplanes passing overhead.
"Intermod," in its true sense, *occurs within the repeater final amplifier*. When signals mix in one or both repeater amplifiers, sum and difference signals on many different frequencies can be heard. **ANSWER B.**

E4D08 What causes intermodulation in an electronic circuit?
    A. Too little gain.   C. Nonlinear circuits or devices.
    B. Lack of neutralization.   D. Positive feedback.
Intermodulation interference is minimized by a careful selection of linear circuits and components. If that newly-designed transceiver for 2 meters or 440 MHz, or that repaired repeater is experiencing high levels of intermodulation, the problem could be caused by *non-linear circuits, devices, and components* inside the receiver. **ANSWER C.**

E4D06 What is the term for unwanted signals generated by the mixing of two or more signals?
    A. Amplifier desensitization.   C. Adjacent channel interference.
    B. Neutralization.   D. Intermodulation interference.
Repeaters are fun to operate through, but a real headache to keep on the air with a clean signal. The problem is *intermodulation* caused by other close-proximity transmitters. What you may hear are *two signals coming in at once*, and this is the effect commonly called "intermod." **ANSWER D.**

E7C10 Which of the following filters would be the best choice for use in a 2 meter repeater duplexer?
    A. A crystal filter.   C. A DSP filter.
    B. A cavity filter.   D. An L-C filter.
For a 2-meter repeater duplexer system, we need big filters. We call them *cavity filters* because they are hollowed out to form extremely narrow-band tuned resonant circuits.
ANSWER B.

4-Cavity Duplexer for 2-Meter Band
Courtesy of WACOM Products, Inc.

E4D04 Which of the following may reduce or eliminate intermodulation interference in a repeater caused by another transmitter operating in close proximity?
A. A band-pass filter in the feed line between the transmitter and receiver.
B. A properly terminated circulator at the output of the transmitter.
C. A Class C final amplifier.
D. A Class D final amplifier.

Your ham friends who regularly trek to mountaintops to maintain ham radio VHF/UHF repeaters deserve your organization's major support. Mountaintop repeaters are located in the worst RF environment possible – in the midst of hundreds of nearby signals that may desensitize the receiver, or create "gar-bage" on the input. Everything was fine when your team left the hill, but a new land-mobile radio system on the same tower is now causing *intermodulation* interference. One *solution is to install a terminated circulator* or ferrite isolator between your own transmitter and duplexer, which helps zero in on your own receive frequency and cancel out any other off-channel frequencies that seem to creep in only when your equipment is on transmit. A spectrum analyzer is an essential piece of equipment for mountaintop repeater maintenance. **ANSWER B.**

E4E08 What type of signal is picked up by electrical wiring near a radio antenna?
A. A common-mode signal at the frequency of the radio transmitter.
B. An electrical-sparking signal.
C. A differential-mode signal at the AC power line frequency.
D. Harmonics of the AC power line frequency.

Electrical wiring within your shack will many times pick up *common mode radio signals* coming off your nearby transmitting antenna. Make sure your radio is well grounded and always suspect interference coming directly off the antenna when you are transmitting. **ANSWER A.**

E7B19 What is a klystron?
A. A high speed multivibrator.
B. An electron-coupled oscillator utilizing a pentode vacuum tube.
C. An oscillator utilizing ceramic elements to achieve stability.
D. A VHF, UHF, or microwave vacuum tube that uses velocity modulation.

The *Klystron* was originally found mainly in RADARs, but a few have been adapted by ham operators for specific microwave frequency use. *Velocity modulation* refers to how this *high-powered tube* is modulated – electrons are beamed down a precisely-tuned tube which has tiny openings that may speed up or slow down the electrons when modulation voltages are applied to the openings. This causes the electron velocity flow to vary. **ANSWER D.**

HERE ARE SOME WEB ADDRESSES OF USEFUL SITES RELATED TO MODULATE YOUR TRANSMITTERS:
www.k8zt.com/racg/Radio_Amateur_%20Conversation_Guide.pdf
www.dx-code.org/
www.qrz.com/
hamcall.net/call
www.dxzone.com/catalog/DX_Resources/Callsigns/

Amps & Power Supplies

E1F03 Under what circumstances may a dealer sell an external RF power amplifier capable of operation below 144 MHz if it has not been granted FCC certification?

A. It was purchased in used condition from an amateur operator and is sold to another amateur operator for use at that operator's station.

B. The equipment dealer assembled it from a kit.

C. It was imported from a manufacturer in a country that does not require certification of RF power amplifiers.

D. It was imported from a manufacturer in another country, and it was certificated by that country's government.

Amplifiers for frequencies below 144 MHz could be *purchased in used condition* by an equipment dealer and *then sold by that dealer to another amateur operator*. This rule prevents the sale of amplifiers really intended for CB use that are constructed without the FCC's blessing. [97.315] **ANSWER A.**

E1F11 Which of the following best describes one of the standards that must be met by an external RF power amplifier if it is to qualify for a grant of FCC certification?

A. It must produce full legal output when driven by not more than 5 watts of mean RF input power.

B. It must be capable of external RF switching between its input and output networks.

C. It must exhibit a gain of 0 dB or less over its full output range.

D. It must satisfy the FCC's spurious emission standards when operated at the lesser of 1500 watts, or its full output power.

Recently, the FCC revised its policy on RF amplifiers. In order for an RF power amplifier to be granted FCC certification, *the amp must meet tight spurious emission standards* when it is operating at full power. Tube amplifiers with large band switching filter networks meet the specs more easily than smaller all-solid-state amps. This may be why the solid state amplifiers are still relatively expensive compared to tube amps. [97.317] **ANSWER D.**

This little 2-meter linear power amplifier meets stringent FCC spurious emission standards.

E1B11 What is the permitted mean power of any spurious emission relative to the mean power of the fundamental emission from a station transmitter or external RF amplifier installed after January 1, 2003, and transmitting on a frequency below 30 MHZ?
A. At least 43 dB below. C. At least 63 dB below.
B. At least 53 dB below. D. At least 73 dB below.

Down on high frequency (below 30 MHz), the Federal Communications Commission requires any spurious emission to be at least 43 dB below the mean power output of that amplifier. Amplifiers contain both harmonic and band pass filters to reduce *spurious emissions down at least 43 dB*. **ANSWER A.**

E8D03 What input-amplitude parameter is valuable in evaluating the signal-handling capability of a Class A amplifier?
A. Peak voltage.
B. RMS voltage.
C. Average power.
D. Resting voltage.

In a *Class A amplifier, peak voltage* provides a double-check on the capability and linearity of that amplifier. Many times, distortion occurs when the amplifier is handling a peak voltage. **ANSWER A.**

E7B04 Where on the load line of a Class A common emitter amplifier would bias normally be set?
A. Approximately half-way between saturation and cutoff.
B. Where the load line intersects the voltage axis.
C. At a point where the bias resistor equals the load resistor.
D. At a point where the load line intersects the zero bias current curve.

The *Class A amplifier* offers the greatest linearity, but offers only marginal efficiency, usually less then 25%. We normally find the Class A amplifier as a "small signal" amplifier, sometimes used in modulation circuits where perfect linearity is required. The input signal continuously varies the input current over the complete 360 degrees of the sine-wave signal, with *bias normally set approximately half-way between saturation and cutoff*. **ANSWER A.**

E7B01 **For what portion of a signal cycle does a Class AB amplifier operate?**
  A. More than 180 degrees but less than 360 degrees.
  B. Exactly 180 degrees.
  C. The entire cycle.
  D. Less than 180 degrees.

The *Class AB amplifier* has some of the properties of both the Class A and the Class B amplifier. It has output for *more than 180 degrees* of the cycle, but *less than 360 degrees* of the cycle. **ANSWER A.**

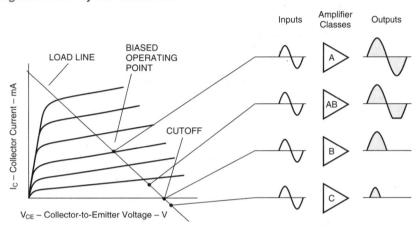

Various Classes of Transistorized Amplifiers

E7B06 **Which of the following amplifier types reduces or eliminates even-order harmonics?**
  A. Push-push.   C. Class C.
  B. Push-pull.   D. Class AB.

The Class-B amplifier may operate a pair of matched tubes in "push-pull" configuration, and operates only over one half of an input cycle. In transistorized Class B amplifiers, the circuit may be called "complimentary-symmetry" where NPN and PNP transistors each conduct over half the cycle. The advantage of the *push-pull circuit is the cancellation of even-order harmonics* – second, fourth, etc. In tube-type Class B amplifiers, it is important that you always change both output tubes together, never replacing one tube and leaving the older tube – always replace in pairs. **ANSWER B.**

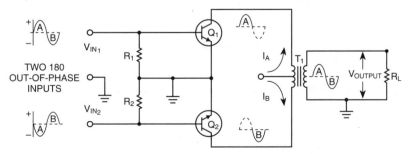

Transistorized Push-Pull Amplifier

E7B18 Which of the following is a characteristic of a grounded-grid amplifier?
A. High power gain.
B. High filament voltage.
C. Low input impedance.
D. Low bandwidth.

The *grounded-grid amplifier* remains one of the most popular methods of achieving legal limit HF power output. Yes, vacuum tubes like the 8877 are still available, and the power supply may only require B+ plate voltage and a tap off the transformer for AC low filament voltage. Another benefit is a 50 ohm *low input impedance*, closely matching the output tuned circuit of your HF transceiver. That tube is literally loafing at a kW, and you'll have a nice clean signal on the air. **ANSWER C.**

E7B08 How can an RF power amplifier be neutralized?
A. By increasing the driving power.
B. By reducing the driving power.
C. By feeding a 180-degree out-of-phase portion of the output back to the input.
D. By feeding an in-phase component of the output back to the input.

We can keep our power-amplified signal clean if the power *amplifier is neutralized* following the instructions in your amplifier manual. Chances are, you will be *feeding a 180-degree out-of-phase portion of the output back to the input* with neutralizing capacitors chosen to obtain the proper amount of feedback. **ANSWER C.**

E7B17 Why are third-order intermodulation distortion products of particular concern in linear power amplifiers?
A. Because they are relatively close in frequency to the desired signal.
B. Because they are relatively far in frequency from the desired signal.
C. Because they invert the sidebands causing distortion.
D. Because they maintain the sidebands, thus causing multiple duplicate signals.

Third-order intermodulation distortion can be a big problem with power amplifiers creating a small amount of in-band *spurious emissions close to your operating frequency*. Third-order IMD will likely elicit ham operators claiming your signal is wide or splattering, especially on 40 meters. **ANSWER A.**

E7B05 What can be done to prevent unwanted oscillations in an RF power amplifier?
A. Tune the stage for maximum SWR.
B. Tune both the input and output for maximum power.
C. Install parasitic suppressors and/or neutralize the stage.
D. Use a phase inverter in the output filter.

As a new Extra soon, you will likely be operating with a tower and beam tied into a power amplifier to boost your transceiver's 100 watt output signal. But cautiously work yourself into high-power amplifier operation. Big amps may cause interference to your home electronics, and big amps need to be tuned precisely to minimize parasitic oscillations. When viewed on a spectrum analyzer, parasitic oscillations look like a forest of trees surrounding your transmitted signal. *In tube amplifiers, RF chokes in the tube plate circuit, and small-value bypass and suppressor capacitors in the grid circuit, may help reduce unwanted oscillations.* With tube amplifiers, step-by-step neutralization procedures are detailed in the instruction book. If the amplifier is new, you likely won't need to go through the neutralization process until you ultimately change out the final amplifier tube. **ANSWER C.**

E7B07 Which of the following is a likely result when a Class C amplifier is used to amplify a single-sideband phone signal?
  A. Reduced intermodulation products.
  B. Increased overall intelligibility.
  C. Signal inversion.
  D. Signal distortion and excessive bandwidth.

The Class AB amplifier will take an input signal and cause a field effect transistor or bipolar transistor to operate near cutoff, but not into cutoff. This will lead to a nearly linear reproduction of the input signal and using Class AB is common with careful attention to setting the bias just above the point where the signal base current goes into cutoff at zero. Once the transistor is biased beyond cutoff, efficiency continues to climb but at the cost of distortion and excessive bandwidth. This is why satellite operators using SSB caution against the use of non-linear FM-only solid state power amplifiers. The Class C amp is fine for data and FM, but for single sideband voice emissions, the *Class C amplifier* is not operating within the linear portion of its operation, *causing the SSB voice signal to sound harsh, distorted, and occupying excessive bandwidth.* **ANSWER D.**

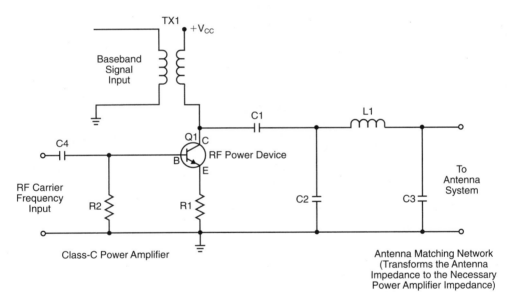

Class-C power amplifier-modulator being modulated with baseband signal.
Source: *Basic Communications Electronics*, Hudson & Luecke, © 1999 Master Publishing, Inc., Niles, IL

E7B02 What is a Class D amplifier?
  A. A type of amplifier that uses switching technology to achieve high efficiency.
  B. A low power amplifier using a differential amplifier for improved linearity.
  C. An amplifier using drift-mode FETs for high efficiency.
  D. A frequency doubling amplifier.

A less-commonly-known amplifier is called Class D, made up of two matched transistors, driven with enough input power to produce a square wave output. This *Class D amplifier employs switching technology to offer high efficiency* from its balanced twin devices. **ANSWER A.**

E7B03 Which of the following forms the output of a class D amplifier circuit?
A. A low-pass filter to remove switching signal components.
B. A high-pass filter to compensate for low gain at low frequencies.
C. A matched load resistor to prevent damage by switching transients.
D. A temperature-compensated load resistor to improve linearity.

The *Class D amplifier* output may be rich in unwanted switching signal components. A *low-pass filter* allows the square wave to pass on to the next stage while attenuating the unwanted higher frequency switching signal components. **ANSWER A.**

E7B21 Which of the following devices is generally best suited for UHF or microwave power amplifier applications?
A. Field effect transistor.   C. Silicon controlled rectifier.
B. Nuvistor.   D. Triac.

The *field effect transistor (FET)* controls current between the source and drain by voltage on the gate. Field effect transistors made with gallium arsenide (GaAs), an improvement over silicon, are preferred at *VHF and UHF* because of high gain and an extremely low noise factor. These are known as GaAs FETs, found in power amplifier receive preamp circuits. **ANSWER A.**

E7B15 What is one way to prevent thermal runaway in a bipolar transistor amplifier?
A. Neutralization.
B. Select transistors with high beta.
C. Use a resistor in series with the emitter.
D. All of these choices are correct.

UHF and microwave operators who homebrew their own amplifiers are cautious to protect against transistor thermal runaway – adding a fixed emitter resistor may slightly cut down on the transistor voltage gain, controlling transistor current without sacrificing linearity. In tube equipment, it is called cathode degeneration. In your transistorized amplifier, *thermal protection* is provided using degenerative emitter feedback, through the *use of a resistor (R 3 in diagram) in series with the emitter*, to ground. **ANSWER C.**

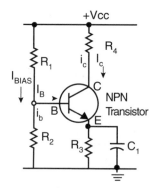

Active Device — Provides Gain

NPN Transistor

Current Gain $= h_{fe} = \dfrac{\Delta\ ic}{\Delta\ ib}$

Passive Device — No Gain

Resistors R_1, R_2, R_3, R_4 and C_1

Δ means "change in"

Small-signal amplifier with an NPN transistor as the active device and resistors and a capacitor as passive devices that set the "no-signal" (DC) operating point.

Source: *Basic Communications Electronics*, Hudson & Luecke, © 1999 Master Publishing, Inc., Niles, IL

E6D09 What devices are commonly used as VHF and UHF parasitic suppressors at the input and output terminals of transistorized HF amplifiers?
   A. Electrolytic capacitors.
   B. Butterworth filters.
   C. Ferrite beads.
   D. Steel-core toroids.

If you look into the modern *VHF/UHF* single-band or dual-band *amplifier*, you'll see many leads dressed on a small *ferrite bead* to minimize parasitics coming down voltage or control lines. **ANSWER C.**

E7B20 What is a parametric amplifier?
   A. A type of bipolar operational amplifier with excellent linearity derived from use of very high voltage on the collector.
   B. A low-noise VHF or UHF amplifier relying on varying reactance for amplification.
   C. A high power amplifier for HF application utilizing the Miller effect to increase gain.
   D. An audio push-pull amplifier using silicon carbide transistors for extremely low noise.

At *VHF, UHF*, and microwave frequencies, amplifier strip-line construction is common to minimize lead length capacitance. A common component within the parametric amplifier is the varactor diode, which behaves much like a variable capacitor, depending on the reverse bias voltage applied. Components are chosen carefully for *low-noise design*, and I personally use a *parametric amplifier* when I am up on a mountaintop operating 10 GHz microwave. **ANSWER B.**

E7B16 What is the effect of intermodulation products in a linear power amplifier?
   A. Transmission of spurious signals.
   B. Creation of parasitic oscillations.
   C. Low efficiency.
   D. All of these choices are correct.

The effects of intermodulation, abbreviated IM, *are spurious signal products* usually traced to power amplifier stages or linear amplifiers with incorrect bias settings. Old time hams called any station with strong IM products "monkey chatter" that may travel up to 200 kHz up and down the band from the main transmitted signal. As a new Extra, with that new power amplifier turned on, start out your operation on extremely low power input to the amp, and suspect IM if another station drops down to your frequency and tells you that your signal is wide, splattering, and you sound like "monkey chatter." With careful tuning and monitoring of amp input drive power from your transceiver, intermodulation can usually be solved without having to completely field strip your station for repair. **ANSWER A.**

E7D01 What is one characteristic of a linear electronic voltage regulator?
  A. It has a ramp voltage as its output.
  B. It eliminates the need for a pass transistor.
  C. The control element duty cycle is proportional to the line or load conditions.
  D. The conduction of a control element is varied to maintain a constant output voltage.

A *linear electronic voltage regulator varies the conduction of a circuit* in direct proportion to variations in the line voltage to, or the load current from, the device. You will find a sophisticated voltage regulation circuit inside your base station power supply, which utilizes a big heavy transformer. **ANSWER D.**

REGULATOR

Voltage Regulator IC
Mounted in Circuit Board

E7D03 What device is typically used as a stable reference voltage in a linear voltage regulator?
  A. A Zener diode.   C. An SCR.
  B. A tunnel diode.   D. A varactor diode.

For *voltage regulation* and a stable reference voltage, *a Zener diode* is found in most ham sets. **ANSWER A.**

ANODE   CATHODE

Here is the schematic symbol of a Zener diode. Since a diode only passes energy in one direction, look for that one-way arrow, plus a "Z" indicating it is a Zener diode. Doesn't that vertical line look like a tiny "Z".

Zener Diode

E7D02 What is one characteristic of a switching electronic voltage regulator?
  A. The resistance of a control element is varied in direct proportion to the line voltage or load current.
  B. It is generally less efficient than a linear regulator.
  C. The control device's duty cycle is controlled to produce a constant average output voltage.
  D. It gives a ramp voltage at its output.

The *switching voltage regulator* actually *switches the control device* completely on or off. **ANSWER C.**

E7D04 Which of the following types of linear voltage regulator usually make the most efficient use of the primary power source?

A. A series current source.
B. A series regulator.
C. A shunt regulator.
D. A shunt current source.

A series regulator normally runs cool. It's *more efficient* than a shunt regulator. Everything passes through the *series regulator*. **ANSWER B.**

E7D05 Which of the following types of linear voltage regulator places a constant load on the unregulated voltage source?

A. A constant current source.
B. A series regulator.
C. A shunt current source.
D. A shunt regulator.

In circuits where the *load on the unregulated input source must be kept constant*, we use the slightly-less-efficient *shunt regulator*. These can sometimes get quite warm, and are usually mounted to the chassis of the equipment. The chassis acts as a heat sink. **ANSWER D.**

E7D06 What is the purpose of Q1 in the circuit shown in Figure E7-3?

A. It provides negative feedback to improve regulation.
B. It provides a constant load for the voltage source.
C. It increases the current-handling capability of the regulator.
D. It provides D1 with current.

This *transistor can handle greater current* than a Zener diode. They sometimes call these transistors "series-pass transistors." **ANSWER C.**

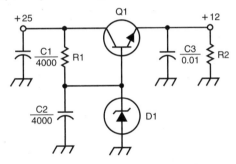

Figure E7-3

E7D07 What is the purpose of C2 in the circuit shown in Figure E7-3?

A. It bypasses hum around D1.
B. It is a brute force filter for the output.
C. To self-resonate at the hum frequency.
D. To provide fixed DC bias for Q1.

C2 provides additional filtering for the ripple frequency. It *bypasses voltage* changes *around* the Zener diode, *D1*, to keep the reference voltage stable. **ANSWER A.**

E7D08 What type of circuit is shown in Figure E7-3?

A. Switching voltage regulator.
B. Grounded emitter amplifier.
C. Linear voltage regulator.
D. Emitter follower.

Components D1 and Q1 give the answer away – *a linear voltage regulator*. This should give you a good, solid, regulated voltage output, even though the load current may be changing. **ANSWER C.**

E7D09 **What is the purpose of C1 in the circuit shown in Figure E7-3?**
    A. It resonates at the ripple frequency.
    B. It provides fixed bias for Q1.
    C. It decouples the output.
    D. It filters the supply voltage.
C1 is an electrolytic *filter* capacitor which removes ripple frequency (60-cycle or 120-cycle) from the *input voltage.* **ANSWER D.**

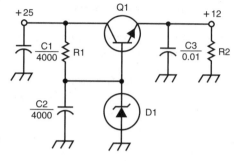

Figure E7-3

E7D10 **What is the purpose of C3 in the circuit shown in Figure E7-3?**
    A. It prevents self-oscillation.
    B. It provides brute force filtering of the output.
    C. It provides fixed bias for Q1.
    D. It clips the peaks of the ripple.
You might see a lot of *C3*-type circuits in your power supply. This is an RF *bypass capacitor*, keeping any stray RF energy from getting into the circuit, which could cause *self-oscillation.* **ANSWER A.**

E7D11 **What is the purpose of R1 in the circuit shown in Figure E7-3?**
    A. It provides a constant load to the voltage source.
    B. It couples hum to D1.
    C. It supplies current to D1.
    D. It bypasses hum around D1.
R1 is connected to the power supply input voltage and *supplies the bias current to* operate *D1* and Q1 at a stable operating point for the voltage regulator output voltage and current desired. **ANSWER C.**

E7D12 **What is the purpose of R2 in the circuit shown in Figure E7-3?**
    A. It provides fixed bias for Q1.   C. It decouples hum from D1.
    B. It provides fixed bias for D1.   D. It provides a constant minimum load for Q1.
R2 connected to the 12-volt output helps maintain voltage regulation by keeping a *constant minimum load* for the transistor Q1, and acts as a bleeder resistor to discharge capacitors after the power supply has been turned off. **ANSWER D.**

E7D13 **What is the purpose of D1 in the circuit shown in Figure E7-3?**
    A. To provide line voltage stabilization.
    B. To provide a voltage reference.
    C. Peak clipping.
    D. Hum filtering.
D1 is a voltage regulator Zener diode, which is used to *provide a voltage reference for Q1.* If the load current increases (R2 gets smaller), the output voltage would tend to decrease. When this voltage change is compared to the reference voltage across D1 by Q1, Q1 adjusts its voltage drop from collector to emitter to raise the output voltage to the designed value with the new load current. **ANSWER B.**

E7D14 What is one purpose of a "bleeder" resistor in a conventional (unregulated) power supply?

A. To cut down on waste heat generated by the power supply.
B. To balance the low-voltage filament windings.
C. To improve output voltage regulation.
D. To boost the amount of output current.

The *bleeder resistor* provides a stable load level in the power supply, and bleeds off current from the charged capacitors when the power supply is shut off. Good *output voltage regulation* is maintained by the bleeder resistor shunt grounded from the charged capacitors to chassis ground. **ANSWER C.**

E7D15 What is the purpose of a "step-start" circuit in a high-voltage power supply?

A. To provide a dual-voltage output for reduced power applications.
B. To compensate for variations of the incoming line voltage.
C. To allow for remote control of the power supply.
D. To allow the filter capacitors to charge gradually.

The modern, high-voltage power supply likely contains a series resistor and capacitor in parallel in the high-voltage plate circuit of the tube in order to buffer the initial in-rush of startup current. This *allows the filter capacitors to slowly build their charges*, minimizing that dreaded arc sound when you turn on a high-voltage supply that has been turned off for several years! Some ham operators bring up high voltage on older equipment with a variable AC transformer. **ANSWER D.**

E7D16 When several electrolytic filter capacitors are connected in series to increase the operating voltage of a power supply filter circuit, why should resistors be connected across each capacitor?

A. To equalize, as much as possible, the voltage drop across each capacitor.
B. To provide a safety bleeder to discharge the capacitors when the supply is off.
C. To provide a minimum load current to reduce voltage excursions at light loads.
D. All of these choices are correct.

Those *resistors* in parallel across power supply electrolytic filter capacitors help *equalize voltage drops* across each of the capacitors, *act as bleeder resistors* when the power supply is turned off, and *add a constant load* if there is minimal current required out of the power supply circuits. **ANSWER D.**

E7D17 What is the primary reason that a high-frequency inverter type high-voltage power supply can be both less expensive and lighter in weight than a conventional power supply?

A. The inverter design does not require any output filtering.
B. It uses a diode bridge rectifier for increased output.
C. The high frequency inverter design uses much smaller transformers and filter components for an equivalent power output.
D. It uses a large power-factor compensation capacitor to create "free power" from the unused portion of the AC cycle.

Conventional, heavy-transformer power supplies that convert home AC power down to 12 volts DC weigh in at about 2 lbs. for every 12 volt DC ampere output. I have plenty of 20 amp power supplies that weigh about 40 lbs. each. These power supplies are characterized by a huge step-down transformer, diodes and big filter capacitors. The best efficiency you could hope for was about 50%. And, ohhhh, would they get warm under load. The new power supply technology is called "switcher," and they are lightweight, small in size, give off little heat, and offer almost 90% efficiency. The output of a switching power supply (whether converting house power down to DC or switching house power up to various high voltage levels) is clean and pure thanks to high-speed field-effect transistors taking the place of that big heavy transformer. New switching power supplies also have minimal EMI (RF interference), and some may even take any detectable emitted spur and let you adjust it out of your favorite listening spot. The *switching power supply* is about the same cost as the old, big-transformer supplies, and the *very small transformer and filter components* easily fit in a relatively small package, either inside or outside the radio equipment. **ANSWER C.**

Receivers with Great Filters

E6E01 What is a crystal lattice filter?
A. A power supply filter made with interlaced quartz crystals.
B. An audio filter made with four quartz crystals that resonate at 1-kHz intervals.
C. A filter with wide bandwidth and shallow skirts made using quartz crystals.
D. A filter with narrow bandwidth and steep skirts made using quartz crystals.

The new worldwide set you purchase will probably have some good SSB and CW filters built in. The filters have a nice *narrow bandwidth and steep skirts*. The filters are made up of *quartz crystals*. You can buy accessory filters for a tighter response, but many times these will make your received signals sound pinched. Unless you do a lot of CW work, to get started, stay with the filters that come with the set. **ANSWER D.**

E6E03 What is one aspect of the piezoelectric effect?
A. Physical deformation of a crystal by the application of a voltage.
B. Mechanical deformation of a crystal by the application of a magnetic field.
C. The generation of electrical energy by the application of light.
D. Reversed conduction states when a P-N junction is exposed to light.

If you *apply a voltage to a quartz crystal*, it will vibrate at a specific frequency. This is the reason crystal oscillators are so accurate. They oscillate at the frequency, or harmonics of the frequency, of the crystal, and continue to do so with great accuracy unless the temperature changes outside design limits. **ANSWER A.**

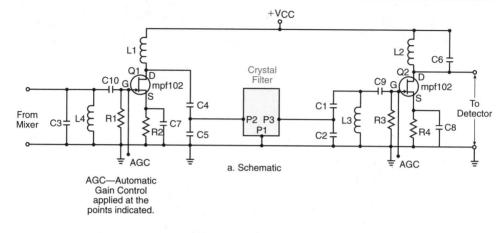

a. Schematic

AGC—Automatic
Gain Control
applied at the
points indicated.

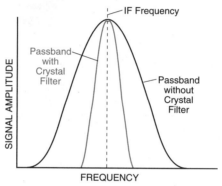

b. Effect of Crystal Filter on Passband

Cascaded IF amplifiers. The active devices are JFETs. The value of most components depends upon the frequency of operation.

Source: *Basic Communications Electronics*, Hudson & Luecke, © 1999 Master Publishing, Inc., Niles, IL

E6E02 Which of the following factors has the greatest effect in helping determine the bandwidth and response shape of a crystal ladder filter?

A. The relative frequencies of the individual crystals.
B. The DC voltage applied to the quartz crystal.
C. The gain of the RF stage preceding the filter.
D. The amplitude of the signals passing through the filter.

The *bandwidth and response of a crystal filter* will relate to the *frequency of the individual crystals* within that filter unit. **ANSWER A.**

E7C05 Which filter type is described as having ripple in the passband and a sharp cutoff?

A. A Butterworth filter.
B. An active LC filter.
C. A passive op-amp filter.
D. A Chebyshev filter.

If you look at the word "*Chebyshev*," it almost looks like it has a *ripple*, doesn't it? **ANSWER D.**

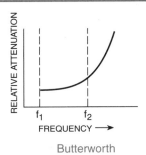

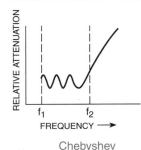

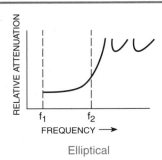

| Butterworth | Chebyshev | Elliptical |

Comparison of Low-Pass Filter Circuits

E7C06 What are the distinguishing features of an elliptical filter?
A. Gradual passband rolloff with minimal stop band ripple.
B. Extremely flat response over its pass band with gradually rounded stop band corners.
C. Extremely sharp cutoff with one or more notches in the stop band.
D. Gradual passband rolloff with extreme stop band ripple.

The *elliptical filter* has an immediate and extremely *sharp cutoff*, with one or more *deep notches* in the stop band. **ANSWER C.**

E4C03 What is the term for the blocking of one FM phone signal by another, stronger FM phone signal?
A. Desensitization.   C. Capture effect.
B. Cross-modulation interference.   D. Frequency discrimination.

A unique property of your FM receiver is the ability to hear only the strongest incoming FM signal. If a distant station is transmitting, and a strong local station comes on the air, you will only hear the local station. *The stronger signal is "capturing" your receiver.* **ANSWER C.**

E7C07 What kind of filter would you use to attenuate an interfering carrier signal while receiving an SSB transmission?
A. A band-pass filter.
B. A notch filter.
C. A Pi-network filter.
D. An all-pass filter.

To *attenuate and block* a specific whistle (or "heterodyne") on an SSB receiver, switch on the notch filter and ever so carefully sweep through the signal until it disappears. Tune slowly! A *notch filter* can require mighty critical adjustment!
ANSWER B. ☞ **Visit: www.nctclearspeech.com**
www.bhinstrumentation.co.uk

E7C08 What kind of digital signal processing audio filter might be used to remove unwanted noise from a received SSB signal?
A. An adaptive filter.
B. A crystal-lattice filter.
C. A Hilbert-transform filter.
D. A phase-inverting filter.

The "*adaptive filter*" found in digital signal processing (DSP) circuits *removes unwanted noise* from incoming SSB signals. **ANSWER A.**

E4D05 What transmitter frequencies would cause an intermodulation-product signal in a receiver tuned to 146.70 MHz when a nearby station transmits on 146.52 MHz?
  A. 146.34 MHz and 146.61 MHz.
  B. 146.88 MHz and 146.34 MHz.
  C. 146.10 MHz and 147.30 MHz.
  D. 173.35 MHz and 139.40 MHz.

To calculate the possible sources of intermodulation interference, we check for the signals that lead to this type of a common problem at base and repeater stations. There are four possible calculations to find the intermodulation interfering frequencies — two where the frequencies are added and two where the frequencies are subtracted. The calculations where the frequencies are subtracted are the only ones used because only these calculations produce frequencies close enough to cause significa nt interference. As the question states, the intermodulation product f_{IMD} is received on 146.70 MHz when a signal at 146.52 MHz is transmitted. The calculations when the frequencies are subtracted are:

$$f_{IMD(2)} = 2f_1 - f_2 \qquad\qquad f_{IMD(4)} = 2f_2 - f_1$$

Rearranging,

$$f_2 = 2f_1 - f_{IMD(2)}$$

$$f_2 = 2(146.52) - 146.70$$

$$f_2 = 146.34 \text{ MHz}$$

Rearranging,

$$2f_2 = f_{IMD(4)} + f_1$$

$$f_2 = \frac{f_{IMD(4)} + f_1}{2}$$

$$f_2 = \frac{146.70 + 146.52}{2}$$

$$f_2 = 146.61 \text{ MHz}$$

ANSWER A.

E4D07 Which of the following describes the most significant effect of an off-frequency signal when it is causing cross-modulation interference to a desired signal?
  A. A large increase in background noise.
  B. A reduction in apparent signal strength.
  C. The desired signal can no longer be heard.
  D. The off-frequency unwanted signal is heard in addition to the desired signal.

You are *listening* to a pal *on the repeater*, and now his *signal is joined by a simultaneous* dispatch of Peter's Perfect Plumbing Service. This is a classic example of *cross-modulation interference* to your friend's signal. **ANSWER D.**

E4C05 What does a value of -174 dBm/Hz represent with regard to the noise floor of a receiver?
  A. The minimum detectable signal as a function of receive frequency.
  B. The theoretical noise at the input of a perfect receiver at room temperature.
  C. The noise figure of a 1 Hz bandwidth receiver.
  D. The galactic noise contribution to minimum detectable signal.

Noise occurs anywhere objects are warmed above absolute zero. If you had a perfect receiver with no internal noise, it would still detect noise when connected to an antenna. *At room temperature, that noise level is about -174 dBm* for every 1 Hz of bandwidth. In addition to the natural noise, man-made noise also occurs, and it is much more prevalent at some frequencies than others. On HF bands, man-made noise is so common that a receiver with a very good noise floor is not nearly as important as it would be for a receiver used at 10 GHz trying to detect an EME signal. **ANSWER B.**

E4C06 A CW receiver with the AGC off has an equivalent input noise power density of -174 dBm/Hz. What would be the level of an unmodulated carrier input to this receiver that would yield an audio output SNR of 0 dB in a 400 Hz noise bandwidth?
  A. 174 dBm.
  B. -164 dBm.
  C. -155 dBm.
  D. -148 dBm.

The weakest signal a receiver can receive is one that is just above the total of all noise in the receiver. -174 dBm/Hz is the background noise the receiver will see at room temperature. Add the noise caused by the receiver's bandwidth to the noise floor of the receiver and you have the total noise. The receiver described in this question matches the -174 dBm/Hz noise of the background, so it is a theoretically-perfect receiver, one with no noise floor of its own. Thus, the weakest signal it can receive is -174 + 10 log (bandwidth) = noise floor = -174 + 10 log (400) + 0 = *-174 + 26 dBm = -148 dBm*. **ANSWER D.**

E4C07 What does the MDS of a receiver represent?
  A. The meter display sensitivity.
  B. The minimum discernible signal.
  C. The multiplex distortion stability.
  D. The maximum detectable spectrum.

The *Minimum Discernable Signal* is tested in the SSB or CW mode, with internal preamps and internal attenuators OUT. Automatic gain control is OFF, and RF gain is full ON. Manufacturers of modern HF equipment have achieved extraordinary low noise floors, with product designs that minimize phase noise. **ANSWER B.**

E4C08 How might lowering the noise figure affect receiver performance?
  A. It would reduce the signal to noise ratio.
  B. It would improve weak signal sensitivity.
  C. It would reduce bandwidth.
  D. It would increase bandwidth.

I know I have a great high frequency receiver when I can turn the volume all the way up with no antenna connected, and hear almost nothing coming out of the receiver. Once I add the antenna, even the weak signals will have a good signal-to-noise ratio. An increase in signal-to-noise ratio is the sign of a quality receiver. Signal-to-noise ratio can be improved either by raising the signal level or by lowering the noise. The *lower the noise floor* of a receiver the *weaker the signals we can detect*, dramatically *improving weak signal sensitivity*. **ANSWER B.**

☞ **Visit: www.flex-radio.com**

E6E05 Which of the following noise figure values is typical of a low-noise UHF preamplifier?

A. 2 dB.

B. -10 dB.

C. 44 dBm.

D. -20 dBm.

Most modern *VHF/UHF* transceivers offer a sensitive internal *preamplifier* to increase weak signals coming in the "front end." Older preamplifiers would actually add a bit of their own unwanted noise, contributing to a noise figure at 3 or 4 dB. Modern preamplifiers using Gallium Arsenide field effect transistors (GaAsFET) can reduce the noise figure well below 2 dB, so when looking for a quality preamplifier circuit, look for the lowest *noise figure below 2 dB.* **ANSWER A.**

E6E06 What characteristics of the MMIC make it a popular choice for VHF through microwave circuits?

A. The ability to retrieve information from a single signal even in the presence of other strong signals..

B. Plate current that is controlled by a control grid.

C. Nearly infinite gain, very high input impedance, and very low output impedance.

D. Controlled gain, low noise figure, and constant input and output impedance over the specified frequency range.

A monolithic microwave integrated circuit (MMIC) looks a bit like a small pill with four leads – two for RF ground, one for RF input, and one for RF output and voltage. *MMICs offer a very low noise figure, good controlled gain, and input and output impedances that remain constant* within their specified operating range. When working with MMICs, be sure you are well grounded, and are working with grounded equipment that will not introduce static electricity. **ANSWER D.**

RF GROUND

RF OUTPUT AND +V_{CC}

RF INPUT (DIAGONAL CUT)

Monolithic Microwave Integrated Circuit
Photo Courtesy of Hewlett Packard Co.

E4C12 What is an undesirable effect of using too wide a filter bandwidth in the IF section of a receiver?

A. Output-offset overshoot.

B. Filter ringing.

C. Thermal-noise distortion.

D. Undesired signals may be heard.

Using an AM 10-kHz IF filter for pulling in 2.3 kHz SSB signals would result in *receiving undesired signals* beyond the normal bandwidth of the desired signal. **ANSWER D.**

E4C10 Which of the following is a desirable amount of selectivity for an amateur RTTY HF receiver?

A. 100 Hz.
B. 300 Hz.
C. 6000 Hz.
D. 2400 Hz.

For *RTTY, 300 Hz is good selectivity.* **ANSWER B.**

E4C11 Which of the following is a desirable amount of selectivity for an amateur SSB phone receiver?

A. 1 kHz.
B. 2.4 kHz.
C. 4.2 kHz.
D. 4.8 kHz.

For *SSB, a 2.4-kHz* filter bandwidth offers good fidelity audio. **ANSWER B.**

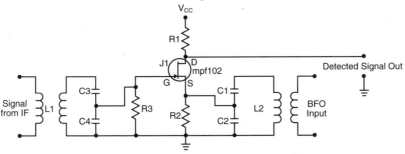

Product detector circuit used for SSB.

E4C13 How does a narrow-band roofing filter affect receiver performance?

A. It improves sensitivity by reducing front end noise.
B. It improves intelligibility by using low Q circuitry to reduce ringing.
C. It improves dynamic range by attenuating strong signals near the receive frequency.
D. All of these choices are correct.

Roofing filters in the IF minimize contesting receiver "swamping" when you have another ham nearby chatting with his buddy, running a kilowatt, well up the band. IF stage roofing filters, like automatically-selected 15 kHz, 6 kHz, and 3 kHz filters, may feature 4-pole, fundamental-mode monolithic crystal filter design to produce an ideal shape factor. The filters usually are positioned after the first mixer and significantly improve the 3rd-order intercept point performance for all stages that follow. I appreciate the performance of roofing filters on 40 meters early in the morning when a nearby shortwave station numbs my receiver. *Roofing filters* eliminate this problem thanks to an *improvement in the rig's dynamic range.* **ANSWER C.**

E4C02 Which of the following portions of a receiver can be effective in eliminating image signal interference?

A. A front-end filter or pre-selector.
B. A narrow IF filter.
C. A notch filter.
D. A properly adjusted product detector.

An effective means of *eliminating image signal interference* is with a *front end filter, or pre-selector filter network.* These filters are much wider than typical IF filters and roofing filters. Also, a higher IF also makes the image further separated from the incoming signal, easing the front end filter requirements. The front end filters are electrically close to the antenna input jack, as opposed to the IF and roofing filters, which are in the middle of the receiver section. The front end filters must be band switched when there is a band change. **ANSWER A.**

E4C09 Which of the following choices is a good reason for selecting a high frequency for the design of the IF in a conventional HF or VHF communications receiver?
A. Fewer components in the receiver.
B. Reduced drift.
C. Easier for front-end circuitry to eliminate image responses.
D. Improved receiver noise figure.

The intermediate frequency (IF) stage in a conventional receiver is where much of the amplification occurs. Selecting a high frequency for the IF allows for improved selectivity with minimum noise. This allows *easier front end circuitry to eliminate image responses* from nearby strong signals. **ANSWER C.**

E4C14 On which of the following frequencies might a signal be transmitting which is generating a spurious image signal in a receiver tuned to 14.300 MHz and which uses a 455 kHz IF frequency?
A. 13.845 MHz.
B. 14.755 MHz.
C. 14.445 MHz.
D. 15.210 MHz.

An image frequency of an incoming desired signal may be plus or minus two times the IF. In this example, the incoming signal is at 14.300 MHz, and the local oscillator is at 14.755 MHz. The incoming 14.300 MHz mixes with 14.755 MHz, producing a different intermediate frequency of 455 kHz and so does an incoming frequency of 15.210 MHz. (15.210 - 14.300 equals 455 kHz, twice.) 15.210 MHz would be the image frequency. Various technologies, including good RF input tuned circuits, can minimize the image frequency thus minimizing the problem. Again, *14.300 MHz + 0.455 + 0.455 = 15.210 MHz*. You double the IF to come up with the correct answer. **ANSWER D.**

E6E12 What is a "Jones filter" as used as part of a HF receiver IF stage?
A. An automatic notch filter.
B. A variable bandwidth crystal lattice filter.
C. A special filter that emphasizes image responses.
D. A filter that removes impulse noise.

The "Jones filter" is a variable-bandwidth crystal filter network, where bandwidth may be changed yet the center frequency remains constant. This filter network allows narrow bandwidth on one side of center, and wide bandwidth on the other side. Shunt varactors operate much like a voltage-controlled oscillator, allowing for variable bandwidth selection. This filter system is found in TenTec equipment, and carries a U.S. patent! *"Variable bandwidth"* is how we remember the new *"Jones filter."* **ANSWER B.**

E4C15 What is the primary source of noise that can be heard from an HF receiver with an antenna connected?
A. Detector noise.
B. Induction motor noise.
C. Receiver front-end noise.
D. Atmospheric noise.

When you hook up an HF transceiver to your antenna, most of the noise received is *atmospheric noise.* **ANSWER D.**

E4C04 What is the definition of the noise figure of a receiver?
A. The ratio of atmospheric noise to phase noise.
B. The noise bandwidth in Hertz compared to the theoretical bandwidth of a resistive network.
C. The ratio of thermal noise to atmospheric noise.
D. The ratio in dB of the noise generated by the receiver compared to the theoretical minimum noise.

For VHF and UHF weak signal work on SSB and CW, you want to operate a receiver with an extremely low noise floor. The GasFET transistor is found in quality VHF/UHF transceivers for weak signal work because of its extremely low noise floor that allows weak signals to be detected above the receiver's internal noise. A *receiver with a good noise floor* will yield nearly *identical noise* power *when switched between a dummy load and an antenna*, measured 3 dB S + N/N, single tone, at 500 Hz and 2400 Hz. This works out to be a ratio, in dB, of the internal noise generated by the receiver itself compared to the theoretical minimum noise that the best receiver in the world would generate! Up on UHF, where we do a lot of weak signal operation, like Moon bounce, this is an important consideration. But down on High Frequency, atmospheric noise would drown out any slight differences in a receiver's own receiver-generated noise, so it is a less important ratio to consider in that new worldwide HF transceiver! **ANSWER D.**

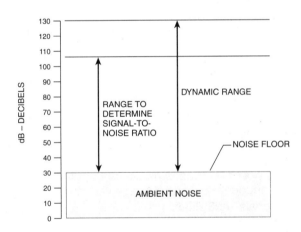

Noise Floor in a Receiver
Source: *Installing and Maintaining Sound Systems*, G. McComb, ©1996, Master Publishing, Inc.

E4D01 What is meant by the blocking dynamic range of a receiver?

    A. The difference in dB between the noise floor and the level of an incoming signal which will cause 1 dB of gain compression.

    B. The minimum difference in dB between the levels of two FM signals which will cause one signal to block the other.

    C. The difference in dB between the noise floor and the third order intercept point.

    D. The minimum difference in dB between two signals which produce third order intermodulation products greater than the noise floor.

When you are working a contest on HF and tune in to a rare station just above the noise floor, you may hear that station request responses to transmit 5 kHz "up." This is contest "split operation", and it leaves the station in the clear without people calling right on top of his ultra weak signal. Powerful stations just a few kilohertz away will sometimes cause the weak station's signal to disappear in the noise, because your receiver is saturated with powerful signals nearby on the band. A receiver with good dynamic range will minimize desensitization when nearby signals blast your front end circuitry. *Blocking dynamic range* is measured in decibels at your *receiver noise floor,* with AGC turned off, and a *nearby signal* that leads to *1 dB of gain compression in the receiver,* causing that weak signal to dip slightly into the noise. Blocking dynamic range tests are more fairly calculated at 100 kHz off from the desired signal. **ANSWER A.**

E4D02 Which of the following describes two problems caused by poor dynamic range in a communications receiver?

    A. Cross-modulation of the desired signal and desensitization from strong adjacent signals.

    B. Oscillator instability requiring frequent retuning and loss of ability to recover the opposite sideband.

    C. Cross-modulation of the desired signal and insufficient audio power to operate the speaker.

    D. Oscillator instability and severe audio distortion of all but the strongest received signals.

In a receiver that does not have good dynamic range, *cross-modulation* of the desired signal can be encountered, and the receiver can be *desensitized* by strong adjacent frequency signals. **ANSWER A.**

E4D09 What is the purpose of the preselector in a communications receiver?

    A. To store often-used frequencies.

    B. To provide a range of AGC time constants.

    C. To increase rejection of unwanted signals.

    D. To allow selection of the optimum RF amplifier device.

You ask why those big contest HF transceivers are so large? On the inside, they may contain high-Q, motor-driven preselector coils that can automatically track, or may be manually adjusted, to your desired receive frequency. These preselector coils may achieve a Q of over 300 and provide steep passband filtering not available with smaller fixed components or older external bandpass boxes. The peak of these optional built-in *preselector* modules may even be adjusted away from your frequency to provide added *isolation from a specific adjacent frequency* transmitting station. The Q is so high that each preselector on each band needs its own micro-motor drive assembly. But for big contesters, the big radio can really cut through the QRM. **ANSWER C.**

E4D10 What does a third-order intercept level of 40 dBm mean with respect to receiver performance?

    A. Signals less than 40 dBm will not generate audible third-order intermodulation products.

    B. The receiver can tolerate signals up to 40 dB above the noise floor without producing third-order intermodulation products.

    C. A pair of 40 dBm signals will theoretically generate a third-order intermodulation product with the same level as the input signals.

    D. A pair of 1 mW input signals will produce a third-order intermodulation product which is 40 dB stronger than the input signal.

If you input two tones that are fHz apart to the receiver input, the mixer will output not only those two tones shifted down to the intermediate frequency but also third-order intermodulation tones at fHz above and below the original tones. At normal input signal levels, these third-order tones are much weaker than the original two tones. But as you increase the amplitude of the original two tones, the *third-order tones* increase in amplitude three times faster than the original tones. If you continue to increase the input tone signal levels, at some point the *signal strength of the input tones and the third-order tones will be equal. This is called the third order intercept point.* A third-order intercept point of 40 dBm means that, if the input tones were 40 dBm, then the third order tones would also be 40 dBm. The 3rd order IP is not achievable in actual practice, as the receiver will saturate long before it reaches that point. **ANSWER C.**

E4D11 Why are third-order intermodulation products created within a receiver of particular interest compared to other products?

    A. The third-order product of two signals which are in the band of interest is also likely to be within the band.

    B. The third-order intercept is much higher than other orders.

    C. Third-order products are an indication of poor image rejection.

    D. Third-order intermodulation produces three products for every input signal within the band of interest.

Your new high-frequency transceiver boasts extraordinary receiver sensitivity, ultra-tight selectivity, and a +30 dBm third order intercept point. Third order intermodulation could occur in the receiver IF or RF stages when a pair of contest stations near you are hammering your "front end" just a few kHz away. Improved equipment designs minimize the *third order "phantom signals"* that are non-desirable weak signal products from these other two stations, *strong enough to interfere with the station you are trying to tune in.* Not only are you hammered by two strong signals of interference, but also artifacts of their clean transmitted signals that may sometimes cover up the desired DX station you are trying to tune in. Physically larger inductors in the band pass filter stage may help, along with low-distortion diodes in the band pass frequency switching circuitry. Monolithic roofing filters, more expensive than overtone mode filters, will offer a better shape factor and are less susceptible to intermodulation distortion within your top quality transceiver's RF and IF stages. If you are considering a contest-capable transceiver to feed into your 100-foot-long boom HF antenna, the multi-thousand-dollar high end HF transceivers will offer you the ultimate in the rejection of third order phantom signals with a distortion-free, high-dynamic-range receiver system. **ANSWER A.**

E4D12 What is the term for the reduction in receiver sensitivity caused by a strong signal near the received frequency?
A. Desensitization.
B. Quieting.
C. Cross-modulation interference.
D. Squelch gain rollback.

If signals mysteriously come in strong then abruptly get weak, and then come back strong again, chances are there is someone nearby transmitting on an adjacent frequency, *desensitizing your receiver*. The best way to check for adjacent signals is through the use of a frequency counter. The very best way to check is with a spectrum analyzer. **ANSWER A.**

E4D13 Which of the following can cause receiver desensitization?
A. Audio gain adjusted too low.
B. Strong adjacent-channel signals.
C. Audio bias adjusted too high.
D. Squelch gain misadjusted.

Adjacent frequency signals may *overload* a receiver's front end causing the receiver to momentarily lose sensitivity. **ANSWER B.**

E4D14 Which of the following is a way to reduce the likelihood of receiver desensitization?
A. Decrease the RF bandwidth of the receiver.
B. Raise the receiver IF frequency.
C. Increase the receiver front end gain.
D. Switch from fast AGC to slow AGC.

If you have a dual-band handheld transceiver, you'll notice that the insides are all enclosed in tiny metal cans. This shielding is necessary to keep the receiver from desensitizing on one band when you are transmitting on another band. Shielding is a good way to minimize internal desensitization. Desensitization occurs when nearby signals get into your receiver and cause the gain to be reduced. By *decreasing the RF bandwidth* of the receiver, you reduce the likelihood of nearby signals affecting your receiver. Only signals that are within the passband of the RF are likely to affect your receive. **ANSWER A.**

E7E10 How does a diode detector function?
A. By rectification and filtering of RF signals.
B. By breakdown of the Zener voltage.
C. By mixing signals with noise in the transition region of the diode.
D. By sensing the change of reactance in the diode with respect to frequency.

Since a *diode* only conducts for half of the AC signal, it *may be used as a rectifier*. By *filtering out the radio frequency energy* after rectification, detection is accomplished – the modulating signal is what remains. **ANSWER A.**

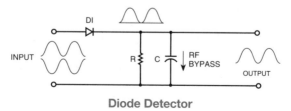

Diode Detector

E7E11 Which of the following types of detector is well suited for demodulating SSB signals?

    A. Discriminator.
    B. Phase detector.
    C. Product detector.
    D. Phase comparator.

A *product detector* is found in *SSB* receivers. It mixes the incoming signal with a beat frequency oscillator signal. The beat frequency oscillator signal is a locally-generated carrier that is mixed with the incoming signal. **ANSWER C.**

E7E06 Why is de-emphasis commonly used in FM communications receivers?

    A. For compatibility with transmitters using phase modulation.
    B. To reduce impulse noise reception.
    C. For higher efficiency.
    D. To remove third-order distortion products.

De-emphasis in an FM receiver rolls off and attenuates the higher modulating frequency response. This compensates for the *FM transmitter using phase modulation*, where the transmitter rolls off the lower frequencies. This provides *compatibility* in the FM communications systems. **ANSWER A.**

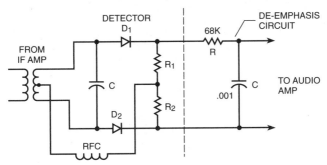

FM De-emphasis Circuit

E7C14 Which of these modes is most affected by non-linear phase response in a receiver IF filter?

    A. Meteor Scatter.
    B. Single-Sideband voice.
    C. Digital.
    D. Video.

Non-linear phase response within a receiver's IF filter is seldom noticed in CW, SSB, or video modes. Unfortunately, complex *digital modes are most affected by non-linear phase response* where individual data pulses incur unwanted shifts of time, disrupting the smooth flow of data within the intermediate frequency filter. Although most filters may incur a small amount of phase distortion, all but data can work through this problem. **ANSWER C.**

E4C01 What is an effect of excessive phase noise in the local oscillator section of a receiver?

    A. It limits the receiver's ability to receive strong signals.
    B. It reduces receiver sensitivity.
    C. It decreases receiver third-order intermodulation distortion dynamic range.
    D. It can cause strong signals on nearby frequencies to interfere with reception of weak signals.

The latest amateur transceivers all boast minimum phase noise in the receiver local oscillator. Any worldwide set with *excessive phase noise* in the L.O. might allow *adjacent frequency signals to interfere* with on-frequency weak signals. **ANSWER D.**

E7E08 What are the principal frequencies that appear at the output of a mixer circuit?

A. Two and four times the original frequency.
B. The sum, difference and square root of the input frequencies.
C. The two input frequencies along with their sum and difference frequencies.
D. 1.414 and 0.707 times the input frequency.

Out of the mixer comes your *original two frequencies, and* the *sum and difference frequencies*. The more elaborate the transceiver, the more mixing stages found in the sets. This helps filter out unwanted signals or phantom signals that could cause interference. **ANSWER C.**

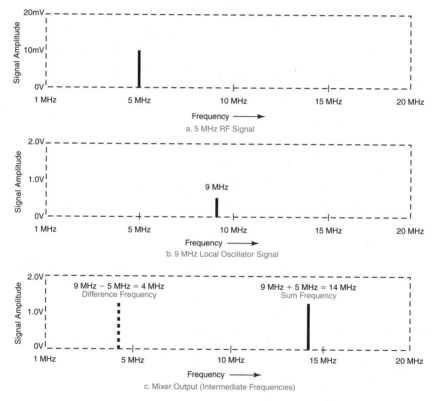

Frequency spectrum plot of mixer signals.

Source: *Basic Communications Electronics*, Hudson & Luecke, © 1999 Master Publishing, Inc., Niles, IL

E7E09 What occurs when an excessive amount of signal energy reaches a mixer circuit?

A. Spurious mixer products are generated.
B. Mixer blanking occurs.
C. Automatic limiting occurs.
D. A beat frequency is generated.

Some VHF power amplifiers may incorporate a hefty pre-amplifier circuit for use on receiving that boosts incoming signal levels. The pre-amp turned on could generate *spurious mixer products as a result of the high signal level*. The spurious products are not really signals that you are trying to receive. If you use a handheld with a power amp, turn the pre-amp off. **ANSWER A.**

E7E14 What is meant by direct conversion when referring to a software defined receiver?

    A. Software is converted from source code to object code during operation of the receiver.

    B. Incoming RF is converted to the IF frequency by rectification to generate the control voltage for a voltage controlled oscillator.

    C. Incoming RF is mixed to "baseband" for analog-to-digital conversion and subsequent processing.

    D. Software is generated in machine language, avoiding the need for compilers.

In a software defined receiver (*SDR*) the *analog to digital conversion begins directly at the input RF stage* where the DSP oscillator translates this signal to baseband. You can see this on high end transceiver spectrum analyzers that are continually displaying a wide range of frequencies without any refresh gap on the signal you have selected. What once took multiple stages in the receiver section can now be accomplished by a software defined receiver, working on your computer. **ANSWER C.**

E7E12 What is a frequency discriminator stage in a FM receiver?

    A. An FM generator circuit.

    B. A circuit for filtering two closely adjacent signals.

    C. An automatic band-switching circuit.

    D. A circuit for detecting FM signals.

Anytime you see the words *frequency discriminator* you know that you are dealing with *signal detection* in an *FM* transceiver. **ANSWER D.**

E4E01 Which of the following types of receiver noise can often be reduced by use of a receiver noise blanker?

    A. Ignition noise.   C. Heterodyne interference.

    B. Broadband white noise.   D. All of these choices are correct.

Going mobile with your HF transceiver? A good *noise blanker will help minimize ignition noise* from recurring spark plug "pops." A manual repetition rate blanker or, better yet, an automatic noise blanker, will synch with the plug "pops" and cancel them almost completely – along with noise from other sources. However, the noise blanker may result in any signal over S-9 sounding distorted. **ANSWER A.**

E4E02 Which of the following types of receiver noise can often be reduced with a DSP noise filter?

    A. Broadband white noise.   C. Power line noise.

    B. Ignition noise.   D. All of these choices are correct.

Most high frequency transceivers now include a digital signal processing filter. This filter will magically subtract broadband white noise, help minimize power line noise, may cancel out a steady frequency heterodyne signal, and help to reduce ignition noise. *All of these may be addressed with a DSP noise filter*. But, you know, I always recommend solving the noise problem at the source, rather than trying to mask it, or subtract it, with DSP. **ANSWER D.**

E4E03 **Which of the following signals might a receiver noise blanker be able to remove from desired signals?**

A. Signals which are constant at all IF levels.
B. Signals which appear across a wide bandwidth.
C. Signals which appear at one IF but not another.
D. Signals which have a sharply peaked frequency distribution.

The *noise blanker* works best on repetitive, correlated, *wide-bandwidth noise* like sparkplug noise, motor noise, and pesky home light dimmers. **ANSWER B.**

E4E07 **How can you determine if line noise interference is being generated within your home?**

A. By checking the power line voltage with a time domain reflectometer.
B. By observing the AC power line waveform with an oscilloscope.
C. By turning off the AC power line main circuit breaker and listening on a battery operated radio.
D. By observing the AC power line voltage with a spectrum analyzer.

Home-office equipment like fax machines or telephones with intercom, fish-tank heaters, and ultrasonic bug repellers all can create whistles and rhythmical pulsing sounds on both high-frequency and VHF/UHF station outputs. To find out whether the noise is being generated by your own home electronics, *shut off your AC power mains and continue listening as you run your rig off of a 12-volt battery source*. If the noise instantly disappears, use your 2-meter handi-talkie and start "sniffing" for the noise source when the power is turned on again. Of course, many home appliances will blink 12:00 after you turn the power back on! **ANSWER C.**

E4E12 **What is one disadvantage of using some types of automatic DSP notch-filters when attempting to copy CW signals?**

A. The DSP filter can remove the desired signal at the same time as it removes interfering signals.
B. Any nearby signal passing through the DSP system will overwhelm the desired signal.
C. Received CW signals will appear to be modulated at the DSP clock frequency.
D. Ringing in the DSP filter will completely remove the spaces between the CW characters.

Digital signal processing (DSP) can relieve your eardrums of constant static roar while waiting for a station to come up on the air. DSP subtracts recurring "hash," allowing only voice, data, and CW tones to pass through to the receiver. The voice may sound slightly pinched with DSP turned on, but you won't hear the static roar between each syllable. A good DSP filter will work magic on CW, as long as the operator is sending at medium to moderate speeds – above 5 wpm. Any *CW signal* sent slower *may be interpreted as an annoying heterodyne by the DSP circuit*, and the circuit will take any continuous steady tone and cancel it – not what you want when copying CW. Just tell the operator to speed up their code sending. The modern DSP notch-filter will recognize the spaces between the dots and dashes, and bring the signal through clearly. **ANSWER A.**

E4E13 What might be the cause of a loud roaring or buzzing AC line interference that comes and goes at intervals?
A. Arcing contacts in a thermostatically controlled device.
B. A defective doorbell or doorbell transformer inside a nearby residence.
C. A malfunctioning illuminated advertising display.
D. All of these choices are correct.

With your new, big, directional HF antenna, taking advantage of your new Extra Class privileges, you may pick up noise in one specific direction, likely when you're aimed at the local shopping center. *All of these* answers *can generate a variety of interference* that may come and go at intervals, just like that flashing neon sign down the street. There is no simple filter to cancel out many of these nuisance noise sources, other than aiming your beam to a point where the noise sources are minimized. **ANSWER D.**

E4E14 What is one type of electrical interference that might be caused by the operation of a nearby personal computer?
A. A loud AC hum in the audio output of your station receiver.
B. A clicking noise at intervals of a few seconds.
C. The appearance of unstable modulated or unmodulated signals at specific frequencies.
D. A whining type noise that continually pulses off and on.

Personal computers and other digital equipment use square-wave signals called *"clocks"* that run at very high frequencies. These clocks can cause problems for radio receivers because they are composed of all the odd harmonics of the fundamental clock frequency. As you tune your receiver up and down the band, you can expect to find those harmonics. They may be *unstable* because clocks are turned on and off as different operations take place in the computer, and they may appear modulated when they are affected by other signals or operations of the computer. If you suspect interference from a digital device, find a frequency where the interference is happening, and then simply turn off the suspected device. If the interference goes away, you've identified the culprit. To reduce the noise from such devices, try moving them farther away from the receiver, or try adding shielding around the device to reduce the level of interference. **ANSWER C.**

E4E10 What is a common characteristic of interference caused by a touch controlled electrical device?
A. The interfering signal sounds like AC hum on an AM receiver or a carrier modulated by 60 Hz hum on a SSB or CW receiver.
B. The interfering signal may drift slowly across the HF spectrum.
C. The interfering signal can be several kHz in width and usually repeats at regular intervals across a HF band.
D. All of these choices are correct.

If you or your neighbor has a *touch lamp, or an electronic light dimmer* control, be prepared to hear *pulse noise* created by these circuits. Sometimes the touch lamp interference will *drift*, sometimes it will have a *hum* on it, and it is *usually wide and repeats itself* up the entire HF band. The touch circuit usually radiates the energy directly through the air, so there is no magic filter to quiet the QRN. Buy your neighbor a new type of lamp control, and ban touch lamps from your home QTH! Oh yes, one more thing – don't be surprised when you operate on 40 meters and these turned off lamp circuits switch on with your nearby transmissions! **ANSWER D.**

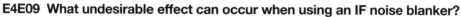

E4E09 What undesirable effect can occur when using an IF noise blanker?

A. Received audio in the speech range might have an echo effect.

B. The audio frequency bandwidth of the received signal might be compressed.

C. Nearby signals may appear to be excessively wide even if they meet emission standards.

D. FM signals can no longer be demodulated.

For mobile applications, a high frequency transceiver will incorporate an IF noise blanker as well as digital noise reduction (DNR), which is part of the transceiver's digital signal processing. If the IF noise blanker is variable in pulse repetition rate, all the better! Most are not. Some transceivers offer blanker on or blanker off with no adjustable amount of noise blanking. Yet others offer variable noise blanker settings. This is better. It is possible that the *IF noise blanker may cause a very strong signal to sound distorted* when the noise blanker is set to maximum. It may also cause strong signals up to 30 kHz away to begin "chattering" your audio circuit on a frequency that is absolutely open. You may be tempted to tell the offending station up frequency that they are "splattering," but the actual problem is not their transmitter (well within specs), but the fact that you left your noise blanker turned on. Turn off your noise blanker when there are strong signals coming in to your mobile or base transceiver! **ANSWER C.**

E4E11 Which of the following is the most likely cause if you are hearing combinations of local AM broadcast signals within one or more of the MF or HF ham bands?

A. The broadcast station is transmitting an over-modulated signal.

B. Nearby corroded metal joints are mixing and re-radiating the broadcast signals.

C. You are receiving sky wave signals from a distant station.

D. Your station receiver IF amplifier stage is defective.

Your modern, high-frequency transceiver employs filter networks, many specifically tuned to minimize powerful double sideband local AM broadcast signals coming in on your lower ham bands. But don't blame the AM broadcaster just yet – *corroded contacts* on your tower or aluminum rain gutters might *re-radiate the broadcast station signals*. This is why many hams with big towers will use multiple insulators in their guy lines to minimize rectification of broadcast band signals due to corroded metal joints. **ANSWER B.**

Oscillate & Synthesize This!

WHEN I GET POSITIVE FEEDBACK, I START TO OSCILLATE!

E7H01 What are three oscillator circuits used in Amateur Radio equipment?
A. Taft, Pierce and negative feedback.
B. Pierce, Fenner and Beane.
C. Taft, Hartley and Pierce.
D. Colpitts, Hartley and Pierce.

Colpitts oscillators have a capacitor, just like C in the name. *Hartley* is tapped, and a *Pierce* oscillator uses a crystal. **ANSWER D.**

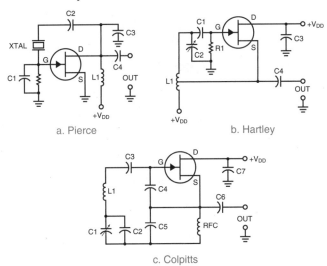

a. Pierce

b. Hartley

c. Colpitts

Three major types of Oscillators

E7H02 What condition must exist for a circuit to oscillate?

A. It must have at least two stages.
B. It must be neutralized.
C. It must have positive feedback with a gain greater than 1.
D. It must have negative feedback sufficient to cancel the input signal.

In order to keep an *oscillator oscillating*, it must have positive feedback sufficient to overcome its natural losses. To keep your kids out there swinging on their swing set, you have to give them a push every so often. That's exactly what *positive feedback* does, but on every cycle. **ANSWER C.**

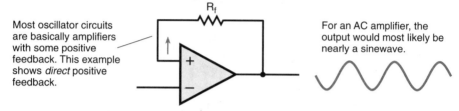

Most oscillator circuits are basically amplifiers with some positive feedback. This example shows *direct* positive feedback.

For an AC amplifier, the output would most likely be nearly a sinewave.

a. Amplifier with Positive Feedback – Oscillator

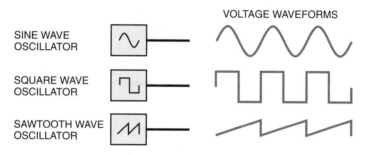

SINE WAVE OSCILLATOR

SQUARE WAVE OSCILLATOR

SAWTOOTH WAVE OSCILLATOR

VOLTAGE WAVEFORMS

b. Oscillator Waveforms

Oscillators – An oscillator is basically an amplifier with positive feedback from output to input.
Source: *Basic Electronics* © 1994, Master Publishing, Inc., Niles, Illinois

E7H03 How is positive feedback supplied in a Hartley oscillator?

A. Through a tapped coil.
B. Through a capacitive divider.
C. Through link coupling.
D. Through a neutralizing capacitor.

Hartley is always tapped. The *tapped coil* provides inductive coupling for positive feedback. Look at the schematic on page 139. **ANSWER A.**

E7H04 How is positive feedback supplied in a Colpitts oscillator?

A. Through a tapped coil.
B. Through link coupling.
C. Through a capacitive divider.
D. Through a neutralizing capacitor.

On the *Colpitts* oscillator, we use a *capacitive divider* to provide feedback. Look at the schematic on page 139. **ANSWER C.**

E7H05 How is positive feedback supplied in a Pierce oscillator?
A. Through a tapped coil.
B. Through link coupling.
C. Through a neutralizing capacitor.
D. Through a quartz crystal.

Every *Pierce* oscillator has a *quartz crystal*, and we use the quartz crystal to obtain positive feedback. Look at the schematic on page 139. **ANSWER D.**

E7H06 Which of the following oscillator circuits are commonly used in VFOs?
A. Pierce and Zener.
B. Colpitts and Hartley.
C. Armstrong and deForest.
D. Negative feedback and balanced feedback.

Since the *Colpitts* oscillator uses a big capacitor in a *variable frequency oscillator* (VFO), it's the most common oscillator circuit for older VFO radios. Newer radios, even though they say they have a VFO, are really digitally-controlled via an optical reader. They just look like they have a large capacitor behind that big tuning dial! **ANSWER B.**

E7H07 What is a magnetron oscillator?
A. An oscillator in which the output is fed back to the input by the magnetic field of a transformer.
B. A crystal oscillator in which variable frequency is obtained by placing the crystal in a strong magnetic field.
C. A UHF or microwave oscillator consisting of a diode vacuum tube with a specially shaped anode, surrounded by an external magnet.
D. A reference standard oscillator in which the oscillations are synchronized by magnetic coupling to a rubidium gas tube.

When you work with X-band and K-band microwave enthusiasts, you may spot a *vacuum tube surrounded by a* big, fat, *horseshoe magnet*. This is the magnetron oscillator, giving the electrons leaving the cathode a spin between the electrodes resulting in the tube oscillating. Keep your screwdrivers several feet away from this powerful horseshoe magnet. Always wear protective lenses around this glass tube. And watch out for high voltages! **ANSWER C.**

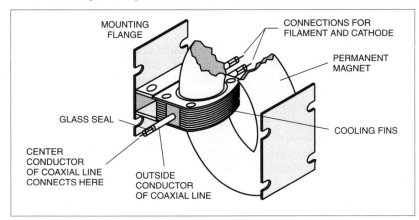

Magnetron

E7H08 What is a Gunn diode oscillator?

A. An oscillator based on the negative resistance properties of properly-doped semiconductors.

B. An oscillator based on the argon gas diode.

C. A highly stable reference oscillator based on the tee-notch principle.

D. A highly stable reference oscillator based on the hot-carrier effect.

Years ago, the Gunn diode oscillator was popular with FM 10 GHz full duplex enthusiasts. We still find Gunn diode oscillators beaming harmless microwave energy at your supermarket door entrance, magically sensing the Doppler shift of someone approaching the doorway, driving a circuit to close a contact to engage the door opener. *Gunn diode oscillators are designed around the negative resistance properties of properly-doped semiconductors.* The Gunn diode is usually mounted in the 10 GHz horn feed point, in a tuned cavity. **ANSWER A.**

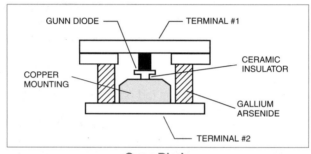

Gunn Diode

E6E09 Which of the following must be done to insure that a crystal oscillator provides the frequency specified by the crystal manufacturer?

A. Provide the crystal with a specified parallel inductance.

B. Provide the crystal with a specified parallel capacitance.

C. Bias the crystal at a specified voltage.

D. Bias the crystal at a specified current.

There is plenty to influence the *precise frequency* of a crystal in the oscillator circuit – crystal impedance, crystal aging, the method of holding the crystal in place inside the can, temperature and, finally, the *crystal's natural parallel capacitance*, usually in the order of 6 pF. **ANSWER B.**

E6E10 What is the equivalent circuit of a quartz crystal?

A. Motional capacitance, motional inductance and loss resistance in series, with a shunt capacitance representing electrode and stray capacitance.

B. Motional capacitance, motional inductance, loss resistance, and a capacitor representing electrode and stray capacitance all in parallel.

C. Motional capacitance, motional inductance, loss resistance, and a capacitor represent electrode and stray capacitance all in series.

D. Motional inductance and loss resistance in series, paralleled with motional capacitance and a capacitor representing electrode and stray capacitance.

Motional capacitance and motional inductance of a quartz crystal refers to the natural series resonance, which is relatively small. This crystal also is part of a series or parallel resonant circuit with a trimmer capacitor or inductor to give it spot-on frequency. And to find the correct answer out of these four choices, commit to memory "*motional capacitance... motional inductance AND loss resistance in series*"....and the following, "shunt capacitance representing electrode and stray capacitance" will fall into place. The job of a quartz crystal is as important as our beating heart. Your question pool committee wants you to fully understand all of the complexities that go into the equivalent circuit of a quartz crystal, and you can read more about this "heart" that is beating within your radio in the **ARRL Handbook for Radio Communications. ANSWER A.**

E7F05 Which of the following is a technique for providing high stability oscillators needed for microwave transmission and reception?

    A. Use a GPS signal reference.
    B. Use a rubidium stabilized reference oscillator.
    C. Use a temperature-controlled high Q dielectric resonator.
    D. All of these choices are correct.

Up at 10 GHz (10,000 MHz) and 24 GHz (24,000 MHz) oscillator stability is of paramount importance. When I am mountain topping with my 10 GHz setup, the sun warming my transmitter can throw me many kHz off frequency! Microwave operators who run beacons will many times use GPS satellite signals for a good stable reference, or may develop a Rubidium standard to keep the oscillator set on the nose. Also, a temperature-controlled high Q dielectric resonator can keep the beacon within a few cycles of being on frequency. *All of these signal references are ways of providing good stability to your microwave system.* **ANSWER D.**

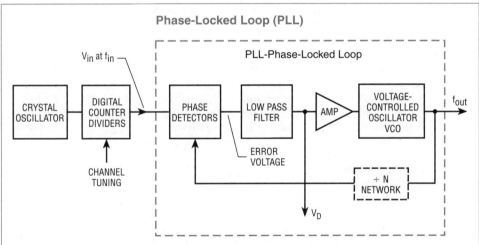

In a phase-locked loop the output frequency of the VCO is locked by phase and frequency to the frequency of the input signal V_{in}.

By using a very accurate crystal controlled source and dividing it down (or multiplying it up), the output frequency, f_{out}, of the PLL will be controlled very accurately and with rigid stability to the input frequency, f_{in}. The phase detector develops an error voltage that is determined by the frequencies and phase of the two inputs. The feedback system wants to reduce the error voltage to zero.

V_D is a voltage proportional to the frequency changes occuring in V_{in}; therefore, the PLL can be used as an FM demodulator.

With divide by N network, f_{out} will be f_{in} multiplied by N.

Channel tuning occurs in the Digital Counter Dividers.

E7H14 What is a phase-locked loop circuit?

A. An electronic servo loop consisting of a ratio detector, reactance modulator, and voltage-controlled oscillator.

B. An electronic circuit also known as a monostable multivibrator.

C. An electronic servo loop consisting of a phase detector, a low-pass filter, a voltage-controlled oscillator, and a stable reference oscillator.

D. An electronic circuit consisting of a precision push-pull amplifier with a differential input.

Few other circuits have changed the course of Amateur Radio equipment as the phase-locked loop (PLL) circuit. There is only one correct answer with the word phase in it, so *look for the word phase when you read phase-locked loop* in the question. **ANSWER C.**

E7H15 Which of these functions can be performed by a phase-locked loop?

A. Wide-band AF and RF power amplification.

B. Comparison of two digital input signals, digital pulse counter.

C. Photovoltaic conversion, optical coupling.

D. Frequency synthesis, FM demodulation.

When I built my first external *frequency synthesizer*, I was able to "synthesize" hundreds of channels with just a couple of crystals. How many of you remember being "rock bound" with an old rig? **ANSWER D.**

E7H09 What type of frequency synthesizer circuit uses a phase accumulator, lookup table, digital to analog converter and a low-pass anti-alias filter?

A. A direct digital synthesizer.

B. A hybrid synthesizer.

C. A phase locked loop synthesizer.

D. A diode-switching matrix synthesizer.

Direct digital synthesis will now allow a single chip to handle multi-loop synthesizer circuits in your radio's local oscillator. Although the DDS does not output a pure sine wave, it offers fractions of a Hertz resolution with an extremely low noise floor. Within the chip is a *phase accumulator, lookup table, digital-to-analog converter, and a low-pass anti-alias filter* with a spectrally clean output. **ANSWER A.**

E7H12 Which of the following is a principal component of a direct digital synthesizer (DDS)?

A. Phase splitter.   C. Chroma demodulator.

B. Hex inverter.   D. Phase accumulator.

Components within that large scale integrated (LSI) direct digital synthesis chip include a crystal oscillator frequency reference, random-access memory, electronic controller, a counter, and a digital-to-analog converter(DAC). One ham radio manufacturer has developed a high spec, ultra-low-noise local oscillator system for their DSS chip that produces a clean first IF signal. The high carrier-to-noise (C/N) ratio of the 200 MHz high-speed digital PLL is the result of a very fast lock time because the local design does not entail the use of a prescaler, but rather locks directly on the first local frequency. This reduces spurious emissions and increases close-in blocking performance. The *phase accumulator* is part of the running step where the phase accumulator is instructed to advance a certain amount by each pulse from the frequency reference. This selects each item in the data table in sequence. The DAC converts this sequence of bits to an analog wave form. **ANSWER D.**

E7H10 What information is contained in the lookup table of a direct digital frequency synthesizer?

    A. The phase relationship between a reference oscillator and the output waveform.

    B. The amplitude values that represent a sine-wave output.

    C. The phase relationship between a voltage-controlled oscillator and the output waveform.

    D. The synthesizer frequency limits and frequency values stored in the radio memories.

Within the DDS, a lookup table will *sample* nearest available *voltage levels* in the digital to analog converter (DAC) *that represent a sine-wave output*. **ANSWER B.** Here is what a block diagram of direct-digital-frequency synthesizer (DDS) looks like:

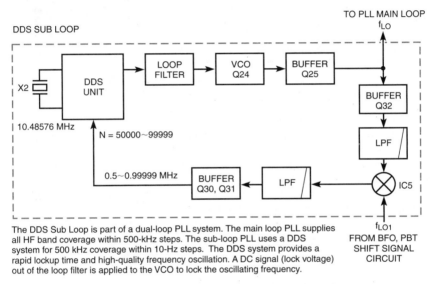

The DDS Sub Loop is part of a dual-loop PLL system. The main loop PLL supplies all HF band coverage within 500-kHz steps. The sub-loop PLL uses a DDS system for 500 kHz coverage within 10-Hz steps. The DDS system provides a rapid lockup time and high-quality frequency oscillation. A DC signal (lock voltage) out of the loop filter is applied to the VCO to lock the oscillating frequency.

DDS Sub Loop of a Transmitter Dual-Loop PLL
Courtesy of ICOM America, Inc.

E7H11 What are the major spectral impurity components of direct digital synthesizers?

    A. Broadband noise.

    B. Digital conversion noise.

    C. Spurious signals at discrete frequencies.

    D. Nyquist limit noise.

Unwanted components of a DDS output are spurs at discrete frequencies – unwanted emissions. These are easier to filter out than broadband noise. However, filtering these numerous spurious signals generated by the direct digital frequency synthesizer is tricky because they are spread over a wide frequency range. Some signals are called "alias signals." These are usually in multiples on each side of the desired frequency and may extend several MHz above and below the desired frequency. Many times a low pass filter may help mitigate the problem with some of these *spurs at discreet frequencies*. **ANSWER C.**

E7H13 What is the capture range of a phase-locked loop circuit?

A. The frequency range over which the circuit can lock.
B. The voltage range over which the circuit can lock.
C. The input impedance range over which the circuit can lock.
D. The range of time it takes the circuit to lock.

When a phase-locked loop circuit drops out of lock it has exceeded the frequency range over which the circuit can effectively lock. *The "capture range" is the frequency range over which a PLL circuit will lock.* **ANSWER A.**

Modern transcievers, like this small hand-held, use PLLs to lock on to the receiving frequency.

E7H17 Why is a phase-locked loop often used as part of a variable frequency synthesizer for receivers and transmitters?

A. It generates FM sidebands.
B. It eliminates the need for a voltage controlled oscillator.
C. It makes it possible for a VFO to have the same degree of frequency stability as a crystal oscillator.
D. It can be used to generate or demodulate SSB signals by quadrature phase synchronization.

Today's modern, high-frequency transceiver with digital direct synthesis has rock solid stability thanks to the 24-bit phase accumulator providing over 16 million points and a frequency resolution of about 0.25 Hz when used with a 5 MHz clock. The output frequency of the *PLL voltage controlled oscillator* is locked by phase and frequency to the frequency of the input signal from the DSS. With a continuous feedback loop to the phase detector, that big transceiver's variable frequency oscillator knob now has the *same degree of stability as a temperature compensated crystal oscillator.* **ANSWER C.**

E7H16 Why is the short-term stability of the reference oscillator important in the design of a phase locked loop (PLL) frequency synthesizer?

A. Any amplitude variations in the reference oscillator signal will prevent the loop from locking to the desired signal.
B. Any phase variations in the reference oscillator signal will produce phase noise in the synthesizer output.
C. Any phase variations in the reference oscillator signal will produce harmonic distortion in the modulating signal.
D. Any amplitude variations in the reference oscillator signal will prevent the loop from changing frequency.

The reference oscillator must have little phase variations because *phase variations will produce phase noise* in the synthesizer output. Excessive phase noise can degrade transmitter and receiver performance. **ANSWER B.**

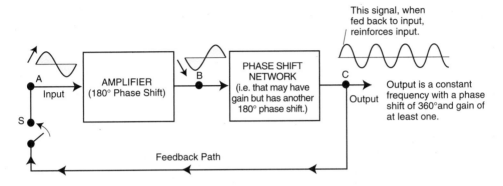

An oscillator outputs a signal of constant frequency.

Source: Basic *Communications Electronics*, Hudson & Luecke, © 1999 Master Publishing, Inc., Niles, IL

E7H18 What are the major spectral impurity components of phase-locked loop synthesizers?

A. Phase noise.
B. Digital conversion noise.
C. Spurious signals at discrete frequencies.
D. Nyquist limit noise.

Phase noise is the nemesis of a PLL synthesizer. **ANSWER A.**

Circuits for Resonance for All!

ELI THE ICE ELMER!

E5A01 What can cause the voltage across reactances in series to be larger than the voltage applied to them?

A. Resonance.
B. Capacitance.
C. Conductance.
D. Resistance.

Coils (having inductance) and capacitors (having capacitance) will be at resonance when the capacitive reactance equals the inductive reactance. *At resonance, large voltages greater than the applied voltage can be present* across these components in the circuit. You can see this with a mobile whip antenna – if something were to touch the whip in transmit, you might see quite a spark! **ANSWER A.**

Equation for Resonant Frequency

X_L = Inductive Reactance

$X_L = 2\pi f L$

X_C = Capacitive Reactance

$X_C = \dfrac{1}{2\pi f C}$

L = Inductance in henrys

C = Capacitance in farads

f = Frequency in hertz

At Resonance $X_L = X_C$

$$2\pi f_r L = \dfrac{1}{2\pi f_r C}$$

Solving for f_r

$$f_r^2 = \dfrac{1}{(2\pi)^2 LC}$$

The Resonant Frequency, f_r, is:

$$f_r = \dfrac{1}{2\pi \sqrt{LC}}$$

On-line calculator site:
www.calculatoredge.com/index.htm

E5A02 What is resonance in an electrical circuit?
A. The highest frequency that will pass current.
B. The lowest frequency that will pass current.
C. The frequency at which the capacitive reactance equals the inductive reactance.
D. The frequency at which the reactive impedance equals the resistive impedance.

At resonance, XL (inductive reactance) equals XC (capacitive reactance). In a mobile series resonant whip antenna, there will be maximum current through the loading coil when the whip is tuned to resonance. On your base station vertical trap antenna or trap beam antenna, parallel resonant circuits offer high impedance, trapping out a specific length of the antenna for a certain frequency. At resonance, the resonant parallel circuit looks like a high impedance to any of your power getting through to the longer length of the antenna system. **ANSWER C.**

E5A03 What is the magnitude of the impedance of a series RLC circuit at resonance?
A. High, as compared to the circuit resistance.
B. Approximately equal to capacitive reactance.
C. Approximately equal to inductive reactance.
D. Approximately equal to circuit resistance.

Did you ever wonder why the serious mobile ham always uses a gigantic loading coil for his series resonant antenna? The bigger the loading coil, the less resistance. *At resonance, the magnitude of the impedance is equal to the circuit resistance.* **ANSWER D.**

E5A04 What is the magnitude of the impedance of a circuit with a resistor, an inductor and a capacitor all in parallel, at resonance?
A. Approximately equal to circuit resistance.
B. Approximately equal to inductive reactance.
C. Low, as compared to the circuit resistance.
D. Approximately equal to capacitive reactance.

Whether the system is parallel or series RLC resonant, the *impedance will always be equal to the circuit resistance.* **ANSWER A.**

E5A05 What is the magnitude of the current at the input of a series RLC circuit as the frequency goes through resonance?
A. Minimum.   C. R/L.
B. Maximum.   D. L/R.

A mobile whip antenna will have *maximum current* in a series RLC circuit *at resonance.* One way you can tell whether or not a mobile whip is working properly is to feel the coil after the transmitter has been shut down. If it's warm, there has been current through the coil and up to the whip tip stinger. **ANSWER B.**

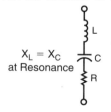

$X_L = X_C$
at Resonance

a. Series

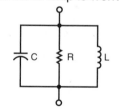

$X_L = X_C$
at Resonance

b. Parallel

Series and Parallel Resonant Circuits

E5A06 What is the magnitude of the circulating current within the components of a parallel LC circuit at resonance?

A. It is at a minimum.

B. It is at a maximum.

C. It equals 1 divided by the quantity 2 times Pi, multiplied by the square root of inductance L multiplied by capacitance C.

D. It equals 2 multiplied by Pi, multiplied by frequency "F," multiplied by inductance "L."

A *parallel circuit at resonance* is similar to a tuned trap in a multi-band trap antenna. At resonance, the trap keeps power from going any further to the longer antenna elements. But be assured there is *maximum circulating current* within the parallel RLC circuit. Now don't get confused with this question – a parallel circuit has maximum current within it, and minimum current going on to the next stage. That's why they call those parallel resonant circuits "traps." **ANSWER B.**

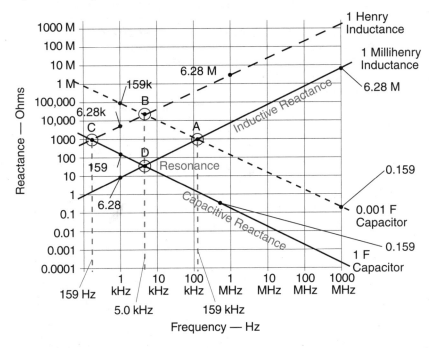

Variation of inductance and capacitive reactance with frequency
(illustration not to exact log-log scale).
Source: Basic *Communications Electronics*, Hudson & Luecke, © 1999 Master Publishing, Inc., Niles, IL

E5A07 What is the magnitude of the current at the input of a parallel RLC circuit at resonance?

A. Minimum.

B. Maximum.

C. R/L.

D. L/R.

The total *current into* and out of a *parallel RLC circuit at resonance is at a minimum*. On old radios, we would tune a plate control to see the current dip at resonance. **ANSWER A.**

E5A08 What is the phase relationship between the current through and the voltage across a series resonant circuit at resonance?
A. The voltage leads the current by 90 degrees.
B. The current leads the voltage by 90 degrees.
C. The voltage and current are in phase.
D. The voltage and current are 180 degrees out of phase.

When a circuit is *at resonance, voltage and current are in phase*. If you remember "ELI the ICE man" it will help you visualize the relationship of voltage and current when they are NOT in resonance. **ANSWER C.**

   ELI = voltage (E) leads current (I) in an inductive (L) circuit.

ICE = current (I) leads voltage (E) in a capacitive (C) circuit.

At *Resonance*: $X_L = X_C$ = voltage and current in phase (neither is leading the other).

E5A09 What is the phase relationship between the current through and the voltage across a parallel resonant circuit at resonance?
A. The voltage leads the current by 90 degrees.
B. The current leads the voltage by 90 degrees.
C. The voltage and current are in phase.
D. The voltage and current are 180 degrees out of phase.

Whether the circuit is series or parallel resonant, at resonance the voltage and current are always in phase. When a mobile whip antenna is tuned for resonance, voltage and current are in phase, and your antenna stainless steel stinger receives all of the power from your transceiver, minus the slight resistance in the network. On that big 3-element trap beam you put up for tri-band operation, the traps are actually parallel resonant circuits, and current WITHIN the trap is at maximum, and only that portion of the antenna BEFORE the trap is radiating. Just remember that voltage and current are in phase at resonance. **ANSWER C.**

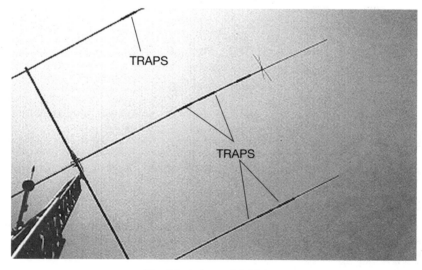

Yagi Antenna with Traps

E5B07 What is the phase angle between the voltage across and the current through a series RLC circuit if X_C is 500 ohms, R is 1 kilohm, and X_L is 250 ohms?

    A. 68.2 degrees with the voltage leading the current.
    B. 14.0 degrees with the voltage leading the current.
    C. 14.0 degrees with the voltage lagging the current.
    D. 68.2 degrees with the voltage lagging the current.

Here is a question on the phase angle between voltage and current in a series RLC circuit. When an applied voltage produces a current in a series RLC circuit, the voltage and current will be out of phase due to the reactance (X_L and X_C) in the circuit. Refer to the schematic below.

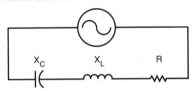

Series RLC Circuit Schematic for Phase Angle Calculations

Here is the formula, memorize it:

$$\text{Tan } \phi = \frac{X}{R}$$

where: ϕ is the phase angle in **degrees**
$X = (X_L - X_C)$ is total reactance in the circuit in **ohms**
R is circuit resistance in **ohms**

Don't panic, it gets easier. Since X_L and X_C are in opposition, we simply subtract one from the other. Now recall ELI the ICE man. Remember him? ELI the ICE man! Say it out loud, "ELI the ICE man." In an inductive (L) circuit, voltage (E) leads current (I) just like in the word "ELI" where E is before I. In a capacitive (C) circuit, current (I) leads voltage (E), just like the word "ICE" where I comes before E. This is an ICE circuit, because X_C is 500 ohms and X_L is only 250 ohms. The total reactance (X) is 500 − 250 = 250. By dividing 1000 (the resistance) into 250 the tangent of ϕ comes out 0.25. We know that a tangent of 0.25 is a *phase angle of 14 degrees, and since the circuit is ICE, voltage lags the current.* (Another way of saying current leads the voltage!) **ANSWER C.**

E5B08 What is the phase angle between the voltage across and the current through a series RLC circuit if X_C is 100 ohms, R is 100 ohms, and X_L is 75 ohms?

    A. 14 degrees with the voltage lagging the current.
    B. 14 degrees with the voltage leading the current.
    C. 76 degrees with the voltage leading the current.
    D. 76 degrees with the voltage lagging the current.

This is more ICE than ELI because X_C is greater than X_L, and the tangent of ϕ of 0.25 gives us a *14-degree phase angle*. Since the *circuit is ICE, voltage (E) is lagging current (I)*. **ANSWER A.**

E5B11 What is the phase angle between the voltage across and the current through a series RLC circuit if X_C is 25 ohms, R is 100 ohms, and X_L is 50 ohms?
    A. 14 degrees with the voltage lagging the current.
    B. 14 degrees with the voltage leading the current.
    C. 76 degrees with the voltage lagging the current.
    D. 76 degrees with the voltage leading the current.

When an applied voltage produces a current in a series RLC circuit, the voltage and current will be out of phase due to the reactance (X_L and X_C) in the circuit. Right off the bat, you can zero in on 2 out of 4 answers by looking to see whether voltage is lagging the current, or voltage is leading the current. Voltage lagging is an ICE (more capacitive) circuit. Voltage leading the current, ELI, is more inductive. Now let's take a look at this problem where X_C is 25 ohms, and X_L is 50 ohms, and R is 100 ohms. The reactance (X) is 50 − 25 = 25. Simple subtraction. Easy stuff. We find the tangent by dividing 100 (the resistance) into 25 = 0.25. From our trig table stored in our brain, the tangent of 14 degrees is 0.25. This gives us a memorized *phase angle of 14 degrees, and voltage is leading the current (ELI)*. Since present and past test questions either ended up with .75 or .25, just keep in mind that .75 = 37 degrees, and .25 = 14 degrees. It looks to me that these series RLC circuits always end up at 14 degrees with you calculating ELI or ICE. Now that wasn't so tough, was it? **ANSWER B.**

Vector Addition

Here is more detail. In an ac circuit, when calculating the impedance of the circuit, the reactance and resistance must be added vectorially rather than algebraically. This vector addition can be understood best by looking at the following diagram:

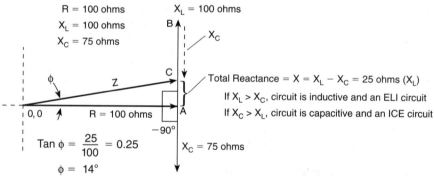

The resistance vector (of a size proportional to its value) is plotted horizontally (along x axis). Any reactance vector (using same scale as for resistance value) is plotted at right angles (90) to the resis-tance vector at the tip of the resistance vector (Point A). Inductive reactance (X_L) would be plotted at +90, while capacitive reactance (X_C) would be plotted at −90. The total reactance is determined by plotting the X_C vector from the tip (Point B) of the X_L vector. The result is Point C on the X_L vector. The total impedance vector, Z, is drawn from the origin (tail of R vector)

to point C. The triangle formed is a right triangle; therefore, by trigonometry and the Pythagorean theorem:

$$Z = \sqrt{R^2 + (X_L - X_C)^2} = \sqrt{100^2 + 25^2}$$
$$= \sqrt{10625} = 103.08 \text{ ohms}$$

and:

$$\text{Tan } \phi = \frac{X}{R} = \frac{25}{100} = 0.25 \quad \phi = 14°$$

The tangent of the angle ϕ is found by dividing the value of the side opposite (X) vector by the value of the side adjacent (R) vector. ϕ is the phase angle of the impedance.

E5B09 What is the relationship between the current through a capacitor and the voltage across a capacitor?
    A. Voltage and current are in phase.
    B. Voltage and current are 180 degrees out of phase.
    C. Voltage leads current by 90 degrees.
    D. Current leads voltage by 90 degrees.
A *capacitor is an "ICE" circuit where current leads the voltage*, as you see in "ICE" where the letter "I" is leading the letter "E." **ANSWER D.**

E5B10 What is the relationship between the current through an inductor and the voltage across an inductor?
    A. Voltage leads current by 90 degrees.
    B. Current leads voltage by 90 degrees.
    C. Voltage and current are 180 degrees out of phase.
    D. Voltage and current are in phase.
This is an *"ELI" condition where voltage leads the current* as we see in "ELI" where the letter "E" is ahead of the letter "I." **ANSWER A.**

E5B12 What is the phase angle between the voltage across and the current through a series RLC circuit if X_C is 75 ohms, R is 100 ohms, and X_L is 50 ohms?
    A. 76 degrees with the voltage lagging the current.
    B. 14 degrees with the voltage leading the current.
    C. 14 degrees with the voltage lagging the current.
    D. 76 degrees with the voltage leading the current.
Here is another ICE circuit, coming out at *14 degrees. Voltage is lagging current (ICE).* X is $75 - 50 = 25$, and tangent of $\phi = 0.25$. **ANSWER C.**

E5B13 What is the phase angle between the voltage across and the current through a series RLC circuit if X_C is 250 ohms, R is 1 kilohm, and X_L is 500 ohms?
    A. 81.47 degrees with the voltage lagging the current.
    B. 81.47 degrees with the voltage leading the current.
    C. 14.04 degrees with the voltage lagging the current.
    D. 14.04 degrees with the voltage leading the current.
Here is a Mr. ELI circuit, coming out at *14 degrees*, where *voltage leads the current*. X is $500 - 250$, and the tangent ϕ is $250 \div 1000 = 0.25$. **ANSWER D.**

E5C13 What coordinate system is often used to display the resistive, inductive, and/or capacitive reactance components of an impedance?
    A. Maidenhead grid.   C. Elliptical coordinates.
    B. Faraday grid.   D. Rectangular coordinates.
The *coordinate system* to display resistive, inductive, and/or capacitive reactance components of an impedance is *rectangular coordinates*. It represents each component reactance or resistance as a vector whose length represents the magnitude of the reactance or resistance. **ANSWER D.**

E5C09 When using rectangular coordinates to graph the impedance of a circuit, what does the horizontal axis represent?
A. Resistive component.
B. Reactive component.
C. The sum of the reactive and resistive components.
D. The difference between the resistive and reactive components.

The horizontal axis represents the resistive component of the impedance. When the impedance is in a circuit and voltage is applied, there will be a current, and the voltage drop across the *resistance is plotted on the horizontal axis.* **ANSWER A.**

Representing an AC Quantity

Rectangular Coordinates

A quantity can be represented by points that are measured on an X-Y plane with an X and Y axis perpendicular to each other. The point where the axes cross is the origin, the points on the X and Y axis are called coordinates, and the magnitude is represented by the length of a vector from the origin to the unique point defined by the *rectangular coordinates.*

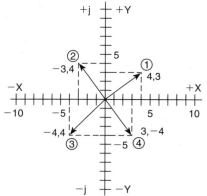

To be able to identify ac quantities and their phase angle, the X axis is called the real axis and the Y axis is called the imaginary axis. There are positive and negative real axis coordinates and positive and negative imaginary axis coordinates. The imaginary axis coordinates have a **j** operator to identify that they are imaginary coordinates.

In rectangular coordinates:
Vector 1 is 4 + j3
Vector 2 is −3 + j4
Vector 3 is −4 − j4
Vector 4 is 3 − j4

Polar Coordinates

A quantity can be represented by a vector starting at an origin, the length of which is the magnitude of the quantity, and rotated from a zero axis through 360°.

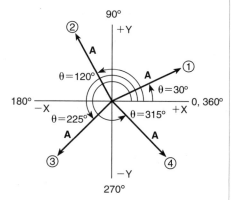

The position of the Vector A can be defined by the angle through which it is rotated. When we do so we are defining the ac quantity in *polar coordinates.*

In polar coordinates:
Vector 1 is A∠30°
Vector 2 is A∠120°
Vector 3 is A∠225°
Vector 4 is A∠315°

E5C22 In rectangular coordinates, what is the impedance of a network consisting of a 10-microhenry inductor in series with a 40-ohm resistor at 500 MHz?

A. 40 + j31,400.   C. 31,400 + j40.

B. 40 − j31,400.   D. 31,400 − j40.

Don't panic, I promise I won't send you back to high school for algebra, trigonometry, and calculus to solve this problem. We have a circuit containing both reactance and resistance, giving us impedance, represented by the letter Z. Impedance has both a resistive (R) and reactive part (X_C or X_L). The parts are represented by vectors that are at right angles (perpendicular) to each other. As a result, $Z = \sqrt{R^2 + X^2}$. Each part forms a leg of a right triangle and contributes to the magnitude of the impedance, which is the hypotenuse of the triangle. You determine the magnitude of the impedance by calculating each part, R and X, and then Z. Let's first calculate the inductive reactance X_L of this question's circuit with the formula for inductive reactance:

E $X_L = 2\pi f L$ Where: L is inductance in **henrys**

f is frequency in **hertz**

π is equal to 3.14

At 500 MHz,

$X_L = 2 \times 3.14 \times 500 \times 10^{+6} \times 10 \times 10^{-6}$

$X_L = 31,400$

The resistance, R, is given as 40 ohms.
Therefore,

$Z = \sqrt{40^2 + 31,400^2} = \sqrt{1600 + 985,960,000} = \sqrt{985,961,600}$

$Z = 31,400.2$ rounded to 31,400

The examination question asks for the answer in rectangular coordinates. This is great news because it allows us to simplify the problem-solving – even to the point of doing the problem in our head. Here's why – rectangular coordinates consist of the 2 parts of the impedance discussed above – the first part is the resistance, and the second part is the reactance. The reactance is preceded by a (+j) or a (-j), which indicates whether the circuit's reactance is inductive with a (+j), or capacitive with a (-j). The resistance is placed first, followed by the reactance: therefore, the impedance of an inductive circuit is $Z = R + jX_L$, and for a capacitive circuit it is $Z = R - jX_C$. The problem has an inductor in series with a 40-ohm resistor. Since it is an inductor, the reactance is inductive and a (+j) precedes it. Therefore, for this major problem, *the easy answer is 40+j*. Now that wasn't so hard, was it? **ANSWER A.**

E5C17 In rectangular coordinates, what is the impedance of a circuit that has an admittance of 5 millisiemens at -30 degrees?

A. 173 − j100 ohms.
B. 200 + j100 ohms.
C. 173 + j100 ohms.
D. 200 − j100 ohms.

$Z = 1/5 \times 10^{-3}\underline{/-30°} = 0.2 \times 10^3\underline{/30°} = 200\underline{/30°}$.

Converting polar coordinates to rectangular gives R = 200 cos30° = 173 ohms, and jX = 200 sin30° = 100 ohms. *The impedance in rectangular coordinates is 173 + j100*, answer C. Be careful of answer A. You will be tempted to put a -j in front of the reactance because the admittance is given initially with a -30° angle. **ANSWER C.**

E5C10 When using rectangular coordinates to graph the impedance of a circuit, what does the vertical axis represent?

A. Resistive component.
B. Reactive component.
C. The sum of the reactive and resistive components.
D. The difference between the resistive and reactive components.

When graphing rectangular coordinates, *the vertical axis represents the reactive component of the impedance*. When the impedance is in a circuit and voltage is applied, there will be a current, and the voltage across the reactance is plotted perpendicular to the horizontal axis and parallel to the vertical axis. Remember – horizontal is resistive, and vertical is reactive. **ANSWER B.**

E5C11 What do the two numbers represent that are used to define a point on a graph using rectangular coordinates?

A. The magnitude and phase of the point.
B. The sine and cosine values.
C. The coordinate values along the horizontal and vertical axes.
D. The tangent and cotangent values.

When we see the two numbers defining a point on a rectangular coordinate graph, these two numbers are *the coordinate values along the horizontal and vertical axes*, respectively. The horizontal coordinate is usually given first. **ANSWER C.**

E5C12 If you plot the impedance of a circuit using the rectangular coordinate system and find the impedance point falls on the right side of the graph on the horizontal axis, what do you know about the circuit?

A. It has to be a direct current circuit.
B. It contains resistance and capacitive reactance.
C. It contains resistance and inductive reactance.
D. It is equivalent to a pure resistance.

If you are computing the impedance using rectangular coordinates, and the *impedance point falls to the right side and directly on the horizontal line*, the impedance is equivalent to a *pure resistance*. **ANSWER D.**

E5C19 **Which point on Figure E5-2 best represents that impedance of a series circuit consisting of a 400 ohm resistor and a 38 picofarad capacitor at 14 MHz?**

A. Point 2.   C. Point 5.

B. Point 4.   D. Point 6.

Look at Figure E5-2, and notice it is a rectangular-coordinate system with axes at right angles to each other. The indicated points on the plane indicate X and Y coordinate values. Watch out for Point 8 on +X axis – this is an 8, and not to be confused with an answer "B." In this question, the resistance of *400 ohms* can be found *along the +X axis*; and since the circuit is capacitive, we go down to *300 on the -Y axis and project a line* parallel to the X axis *to intersect at Point 4* a projection from 400 on the +X axis. You can work the problem all the way out, but in this question you can visually spot it at a glance. **ANSWER B.**

To check yourself, solve the formula $X_C = 1/2\pi fC$ as follows:

$$X_C = \frac{1}{2\pi fC} = \frac{1}{6.28 \times 14 \times 10^6 \times 38 \times 10^{-12}}$$

$$= \frac{1}{3.341 \times 10^3 \times 10^{-6}} = 0.2993 \times 10^3 = 299.3 \text{ ohms or 300 ohms}$$

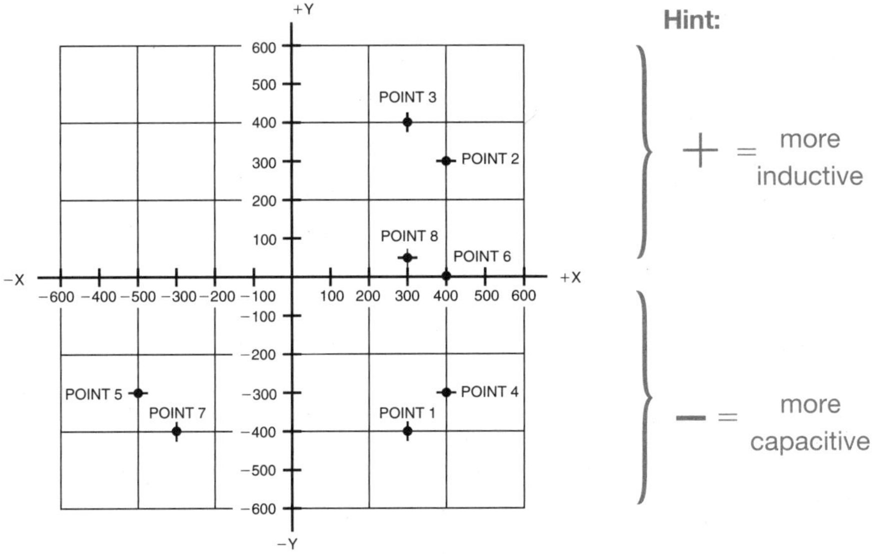

Hint:

$+ =$ more inductive

$- =$ more capacitive

Figure E5-2

E5C20 **Which point in Figure E5-2 best represents the impedance of a series circuit consisting of a 300 ohm resistor and an 18 microhenry inductor at 3.505 MHz?**

A. Point 1.   C. Point 7.

B. Point 3.   D. Point 8.

The inductive reactance at 3.505 MHz is $X_L = 2\pi \times 3.505 \times 10^6 \times 18 \times 10^{-6} = 396.2$ or approximately 400 ohms of inductive reactance. *Follow the +X axis out to 300* for the resistance value. Move up on a line parallel to the *+Y axis* until you intersect the *400 line*. You need to go in a positive Y direction because the circuit is inductive. *Point 3* is the answer. **ANSWER B.**

E5C21 Which point on Figure E5-2 best represents the impedance of a series circuit consisting of a 300 ohm resistor and a 19 picofarad capacitor at 21.200 MHz?

    A. Point 1.   C. Point 7.

    B. Point 3.   D. Point 8.

The capacitive reactance is $X_C = 1/2\pi fC = 1/6.28 \times 21.2 \times 10^6 \times 19 \times 10^{-12}$ $= 1/2.53 \times 10^{-3} = 395$ ohms. We first move to the right along the $+X$ *axis* to the value of the *300-ohm resistor*, and move *DOWN* on a line perpendicular to the X axis (parallel to *the -Y axis*) *to Point 1* at the intersection with the minus 400 line. 395 ohms is considered as 400 ohms. (If we went up, rather than down, we would be incorrect. Going up the $+Y$ axis would mean the circuit is inductive. 19 picofarads makes the circuit capacitive, so we go down.) **ANSWER A.**

E5C23 Which point on Figure E5-2 best represents the impedance of a series circuit consisting of a 300-ohm resistor, a 0.64-microhenry inductor and an 85-picofarad capacitor at 24.900 MHz?

    A. Point 1.   C. Point 5.

    B. Point 3.   D. Point 8.

In this problem we must calculate both inductive and capacitive reactance.

$$X_L = 2\pi \times 24.9 \times 10^6 \times 0.64 \times 10^{-6} = 100 \text{ ohms.}$$

$$X_C = \frac{1}{2\pi \times 24.9 \times 10^6 \times 85 \times 10^{-12}} = \frac{1}{1.33 \times 10^{-2}} = 75 \text{ ohms.}$$

The inductive reactance (100 ohms) minus the capacitive reactance (75 ohms) results in 25 ohms final reactance for this series circuit, which is slightly inductive at 24.9 MHz. Move out to the right to *the 300-ohm resistance line on the $+X$ axis*. Move up slightly to the $+Y$ *axis to find Point 8 at 25 ohms*, which is the correct answer. **ANSWER D.**

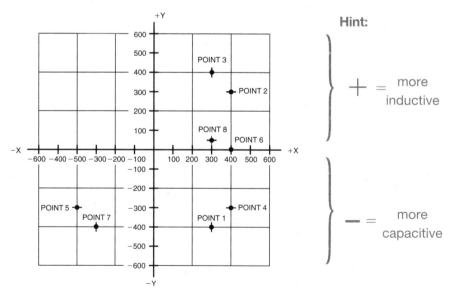

Figure E5-2

E5C14 What coordinate system is often used to display the phase angle of a circuit containing resistance, inductive and/or capacitive reactance?
A. Maidenhead grid.
B. Faraday grid.
C. Elliptical coordinates.
D. Polar coordinates.

To display the impedance of a circuit with its phase angle, you would use a polar coordinate system. *Polar coordinates provides a visual representation of the value of the impedance and its phase angle.* **ANSWER D.**

E5C04 In polar coordinates, what is the impedance of a network consisting of a 400-ohm-reactance capacitor in series with a 300-ohm resistor?
A. 240 ohms at an angle of 36.9 degrees.
B. 240 ohms at an angle of -36.9 degrees.
C. 500 ohms at an angle of 53.1 degrees.
D. 500 ohms at an angle of -53.1 degrees.

Here is a series problem which is easily solved by first calculating the magnitude of the impedance, which equals the square root of resistance squared plus reactance squared. The square root of 300 squared plus 400 squared is the square root of 250,000 or *500 ohms*. Since the circuit is capacitive, the phase angle will have a minus sign in front of it. Remember again, a right triangle with a 3:4:5 relationship of the sides and hypotenuse is an easy one to determine the hypotenuse value. With sides of 300 and 400, the hypotenuse will be 500, with a phase angle whose tangent is 3/4 or 4/3 depending on the relationship of the sides. In this case it's 4/3 and the *angle is -53.1 degrees.* **ANSWER D.**

E5C01 In polar coordinates, what is the impedance of a network consisting of a 100-ohm-reactance inductor in series with a 100-ohm resistor?
A. 121 ohms at an angle of 35 degrees.
B. 141 ohms at an angle of 45 degrees.
C. 161 ohms at an angle of 55 degrees.
D. 181 ohms at an angle of 65 degrees.

This one is relatively easy – impedance equals the square root of the sum of the resistance squared and the inductive reactance squared. Z in polar coordinates is:

$$Z = \sqrt{100^2 + 100^2} \; \underline{/\text{arctan } 100/100}$$

$$Z = \sqrt{20000} \; \underline{/\text{arctan } 1}$$

$$Z = 141.4 \; \underline{/+45°}$$

You might recognize the square root easier if you work in powers of 10. The magnitude of $Z = \sqrt{(1 \times 10^2)^2 + (1 \times 10^2)^2} = \sqrt{2 \times 10^4} = \sqrt{2} \times 10^2 = 1.414 \times 10^2$. You may recognize the $\sqrt{2}$ as 1.414, so *the answer is 141.4.* From just the magnitude, you could have selected the correct answer, or you might have used the phase angle. A right triangle with equal R and X has a *phase angle of 45°* because, as you probably recall, the tangent of 45° is 1. **ANSWER B.**

E5C05 In polar coordinates, what is the impedance of a network consisting of a 400-ohm-reactance inductor in parallel with a 300-ohm resistor?

    A. 240 ohms at an angle of 36.9 degrees.
    B. 240 ohms at an angle of -36.9 degrees.
    C. 500 ohms at an angle of 53.1 degrees.
    D. 500 ohms at an angle of -53.1 degrees.

We know right off the bat that the answer is going to be either A or C because an inductive reactance will have a positive polar coordinate angle. To come up with 240 ohms at an angle of $+36.9°$ as the correct impedance, let's start out by defining the two impedances that are in parallel:

$$Z_1 = 300 \,\underline{/0°} \text{ since it is a resistance in } \textbf{ohms}$$
$$Z_2 = 400 \,\underline{/+90°} \text{ since it is an inductive reactance in } \textbf{ohms}$$

▶ $$Z = \frac{Z_1 \times Z_2}{Z_1 + Z_2} \quad \text{or Special Case: } Z = \frac{RX_L \,\underline{/+90°}}{\sqrt{R^2 + X_L^2} \,\underline{/arctan}}$$

Additions and subtractions are easier in rectangular coordinates, and multiplication and division are easier in polar coordinates.

$$Z = \frac{300 \times 400 \,\underline{/0° + 90°}}{(300 + j0) + (0 + j400)} = \frac{120,000 \underline{/+90°}}{500 \,\underline{/arctan\ 400/300}}$$

$$Z = 240 \,\underline{/+90°-53.1°}$$

$$Z = 240 \,\underline{/36.9°} \text{ ohms}$$

The correct answer is A. We used the 3:4:5 ratio, and both rectangular and polar coordinates. **ANSWER A.**

E5C02 In polar coordinates, what is the impedance of a network consisting of a 100-ohm-reactance inductor, a 100-ohm-reactance capacitor, and a 100-ohm resistor, all connected in series?

    A. 100 ohms at an angle of 90 degrees.
    B. 10 ohms at an angle of 0 degrees.
    C. 10 ohms at an angle of 90 degrees.
    D. 100 ohms at an angle of 0 degrees.

Easy one again – since inductive reactance cancels capacitive reactance, in this circuit with both reactances equal, the final reactance is zero, and all we have left is 100 ohms resistance. Since there is no reactance vector, there is no phase difference, so the phase angle is zero degrees. The resistance vector lies on the 0° axis. One of the easiest ways to check this is to put the impedance in rectangular coordinates: Z = 100 +j100 -j100 = 100 +j0, just *a resistance of 100 ohms lying on the zero axis.* **ANSWER D.**

E5C03 In polar coordinates, what is the impedance of a network consisting of a 300-ohm-reactance capacitor, a 600-ohm-reactance inductor, and a 400-ohm resistor, all connected in series?

    A. 500 ohms at an angle of 37 degrees.
    B. 900 ohms at an angle of 53 degrees.
    C. 400 ohms at an angle of 0 degrees.
    D. 1300 ohms at an angle of 180 degrees.

Even though the answer is requested in polar coordinates, it is best to start out using rectangular coordinates. Z = (400 +j600 -j300) = (400 +j300). Remember, in a right triangle with a 3:4:5 relationship, it is easy to determine the value of the hypotenuse. With sides of 300 and 400, the hypotenuse will be 500, with a phase angle whose tangent is 3/4 or 4/3, depending upon the relationship of the sides. With the resistance side of a right triangle at 400 ohms and the reactance side at 300 ohms, you know that the hypotenuse magnitude of the *impedance is Z = 500*. The angle is a positive one with an arctan of (an angle whose tangent is) 300/400, or 0.75. Grab a trig table or scientific calculator for *an angle of 36.8° rounded to 37°*. **ANSWER A.**

E5C06 In polar coordinates, what is the impedance of a network consisting of a 100-ohm-reactance capacitor in series with a 100-ohm resistor?
    A. 121 ohms at an angle of -25 degrees.
    B. 191 ohms at an angle of -85 degrees.
    C. 161 ohms at an angle of -65 degrees.
    D. 141 ohms at an angle of -45 degrees.
R = 100 ohms and -X_C = 100 ohms. Calculate Z = $\sqrt{100^2 + 100^2}$ = $\sqrt{2 \times 10^4}$. This works out to be the square root of 20,000, which is *141 ohms*. The phase angle will be negative so the correct answer D can be selected without going further. However, the arctan is 100/100 or 1, and from this we know that the *angle is 45°*. **ANSWER D.**

E5C07 In polar coordinates, what is the impedance of a network comprised of a 100-ohm-reactance capacitor in parallel with a 100-ohm resistor?
    A. 31 ohms at an angle of -15 degrees.
    B. 51 ohms at an angle of -25 degrees.
    C. 71 ohms at an angle of -45 degrees.
    D. 91 ohms at an angle of -65 degrees.
Polar coordinates again, but you are given X_C so you don't have to calculate it. First calculate impedance:

$$Z = \frac{100\ \underline{/0°} \times 100\ \underline{/-90°}}{\sqrt{100^2 + 100^2}\ \underline{/\text{arctan} -1}} = \frac{10000\ \underline{/-90°}}{141.4\ \underline{/-45°}}$$

$$Z = 70.7\ \underline{/-45°}$$

You can determine the correct answer as C just from the impedance magnitude, but since the arctan is 1, you know the phase angle is 45°. **ANSWER C.**

E5C16 In polar coordinates, what is the impedance of a circuit that has an admittance of 7.09 millisiemens at 45 degrees?
    A. 5.03 × 10⁻⁶ ohms at an angle of 45 degrees.
    B. 141 ohms at an angle of -45 degrees.
    C. 19,900 ohms at an angle of -45 degrees.
    D. 141 ohms at an angle of 45 degrees.
Admittance is the reciprocal of impedance. Admittance is 1/Z; therefore: Z = 1/A = 1/7.09 × 10⁻³$\underline{/45°}$ = 141$\underline{/-45°}$. Remember, the +45° angle in the denominator becomes minus 45° when divided into 1. **ANSWER B.**

E5C08 In polar coordinates, what is the impedance of a network comprised of a 300-ohm-reactance inductor in series with a 400-ohm resistor?
A. 400 ohms at an angle of 27 degrees.
B. 500 ohms at an angle of 37 degrees.
C. 500 ohms at an angle of 47 degrees.
D. 700 ohms at an angle of 57 degrees.

Easy one here – let's find the magnitude of the impedance and see if the correct answer can be determined just from it. The phase angle will be positive because the circuit is inductive. You don't really have to calculate if you recognize that the right triangle sides have a ratio of 3:4:5. Since resistance is 400 ohms, inductive reactance 300 ohms, *Z will have a magnitude of 500 ohms. The angle is 37°.* You don't need to know more than that. **ANSWER B.**

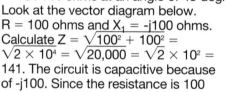

$$\text{Tangent } \Theta = \frac{\text{X Reactance}}{\text{R Resistance}} = \frac{300}{400} = 0.75$$

E5C15 In polar coordinates, what is the impedance of a circuit of 100 -j100 ohms impedance?
A. 141 ohms at an angle of -45 degrees.
B. 100 ohms at an angle of 45 degrees.
C. 100 ohms at an angle of -45 degrees.
D. 141 ohms at an angle of 45 degrees.

Look at the vector diagram below.
$R = 100$ ohms and $X_1 = -j100$ ohms.
Calculate $Z = \sqrt{100^2 + 100^2} =$
$\sqrt{2 \times 10^4} = \sqrt{20,000} = \sqrt{2} \times 10^2 =$
141. The circuit is capacitive because
of -j100. Since the resistance is 100
ohms and the capacitive reactance is
100 ohms, the angle (Θ) has a tangent
of -1; therefore, the angle is a negative
45°. *The impedance in polar coordinates is 141$\angle-45°$.* **ANSWER A.**

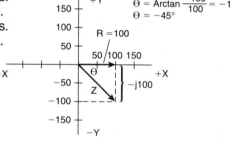

Vector Diagram for E5C15

E5C18 In polar coordinates, what is the impedance of a series circuit consisting of a resistance of 4 ohms, an inductive reactance of 4 ohms, and a capacitive reactance of 1 ohm?
A. 6.4 ohms at an angle of 53 degrees.
B. 5 ohms at an angle of 37 degrees.
C. 5 ohms at an angle of 45 degrees.
D. 10 ohms at an angle of -51 degrees.

Start your calculation by using
rectangular coordinates. $Z = (4 +$
$j4 - j1) = (4 + j3)$. Since the circuit is
inductive, the answer will have positive
degrees. And you will know that a
right triangle with a 3:4:5 relationship
of the sides and hypotenuse is an
easy one to determine the hypotenuse

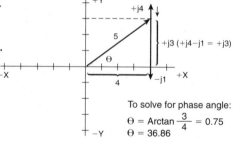

To solve for phase angle:
$$\Theta = \text{Arctan } \frac{3}{4} = 0.75$$
$$\Theta = 36.86$$

Vector Diagram for E5C18

value. With sides 3 and 4, the hypotenuse will be 5, with a phase angle whose tangent is 3/4 or 4/3 depending on the relationship of the sides. *You have the 3 and 4, and answer B has the 5 with no negative sign in front of 37 degrees.* **ANSWER B.**

A Different, Easier, and Even COOLER TRICK!!!
Courtesy of Joseph J. Castanza, Ph.D., AC2FQ

Here's an even easier way to approach all of these Polar Coordinate problems without using the Polar Coordinate conversion function (still devoid of any scary math!!!) in only 475 words! Ready? Go!

If the circuit is in series, simply divide the reactance value by the resistance value, and then hit the arctan (TAN-1) button on the calculator to find the phase angle; there's either an arctan key or, on most modern scientific calculators, you press either a [SHIFT], [FN], [ ] or [FUNCTION] key and then press the regular TAN key. If the reactance is capacitive, then the angle sign becomes negative. We STILL only care about correctly-signed angles – not impedances! Ready? Let's try it!
Back to question E5C04 (p.161): In polar coordinates, what is the impedance of a network consisting of a 400-ohm-reactance capacitor in series with a 300-ohm resistor?
  1. The reactance (capacitive) = 400 ohms, and the resistance = 300 ohms:
  2. It's a series circuit; so: (reactance ÷ resistance) = 400 ÷ 300 = 1.3333
  3. With that 1.3333 result still on the screen, just hit the arctan button (or hit [SHIFT] or [FN] and then hit the [TAN] button) to obtain the phase angle: Angle = 53.13°. We know that this is a negative phase angle because the reactance is capacitive, so, look for the answer that has -53.1 degrees, and that answer is D. 500 ohms at an angle of -53.1 degrees.

Conversely, if the circuit is in parallel (as in question E5C05 on p.162), then we divide the resistance by the reactance, which is the inverse (opposite) of what we did for E5C04 (those crazy, backward parallel circuits!), and then we just plug that result into the arctan function. Try this:
E5C05: In polar coordinates, what is the impedance of a network consisting of a 400-ohm reactance inductor in parallel with a 300-ohm resistor?
  1. The reactance (inductive) = 400 ohms, and the resistance = 300 ohms:
  2. It's a parallel circuit so, this time, we divide the resistance by the reactance (instead of the other way around, as we did in E5C04). So: (resistance ÷ reactance) = 300 ÷ 400 = 0.75.
  3. With that 0.75 result still on the screen, just hit the arctan button (or hit [SHIFT] or [FN] and then hit the [TAN] button) to obtain the angle: Arctan (0.75) = 36.87°. That's a positive 36.9° because the reactance is inductive. The correct answer choice is A. 240 ohms at an angle of 36.9°.

Here are the steps for dealing with all 10 Polar Coordinate questions:
  1. Set your calculator to DEGREE mode and figure out whether the circuit is in parallel or series.
  2. If it's in series, divide the reactance by the resistance. If it's in parallel, divide the resistance by the reactance (remember "those crazy, backward, parallel circuits!" so you know what to do).
  3. With that result still on the screen, press the arctan button (or [SHIFT] and then [TAN]).
  4. Correct the sign of the displayed phase angle for the reactance: if it's capacitive, the angle is negative; if it's inductive, the angle is positive.
  5. Ignore impedance! Just find the answer with the correctly-signed angle and you're done!

THAT'S IT!!! And, guess what? This is all that the Polar Coordinate conversion function was doing for us the entire time! I've checked this for all 10 of the Polar Coordinate questions and it works just fine (with common sense used on E5C02 to realize that the angle is 0° and that there are only 100 ohms of resistance remaining; and on E5C03, which requires subtraction of 300 capacitive ohms from 600 inductive). Try them all for yourself!

E4B15 Which of the following can be used as a relative measurement of the Q for a series-tuned circuit?
- A. The inductance to capacitance ratio.
- B. The frequency shift.
- C. The bandwidth of the circuit's frequency response.
- D. The resonant frequency of the circuit.

In a series-tuned circuit, the higher the Q of a large antenna coil the narrower the bandwidth of frequencies it will pass. Small helical-coil, high-frequency mobile whip antennas have a relatively lower Q, offering good *bandwidth*, but with less efficiency. **ANSWER C.**

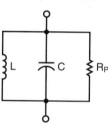

A large loading coil on a mobil whip antenna helps achieve high Q resonance

For tuned circuits the quality factor, Q, is:

Series Resonant Circuit

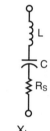

$$X_L = 2\pi f_r L$$

$$X_C = \frac{1}{2\pi f_r C}$$

$X_L = X_C$ at f_r, the resonant frequency

$$Q = \frac{X_L}{R_S} \text{ or } \frac{X_C}{R_S}$$

Parallel Resonant Circuit

$$Q = \frac{R_P}{X_L} \text{ or } \frac{R_P}{X_C}$$

For tuned circuits with Q greater than 10:

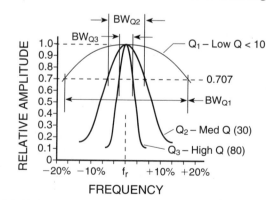

f_r = Resonant frequency in Hz

BW = Half-power Bandwidth ($-$3db)

$$Q = \frac{f_r}{BW}$$

$$\therefore BW = \frac{f_r}{Q}$$

Q is the quality factor for a tuned circuit.
Source: *Advanced Class*, 2ED, G. West, © 1992, 1995, Master Publishing Inc., Niles, IL.

E5A10 What is the half-power bandwidth of a parallel resonant circuit that has a resonant frequency of 1.8 MHz and a Q of 95?

A. 18.9 kHz.　　　　　　　　　　　C. 94.5 kHz.
B. 1.89 kHz.　　　　　　　　　　　D. 9.45 kHz.

Here is a series of questions on the half-power bandwidth of a parallel resonant circuit. There is a good chance one of these question will be asked on your actual Extra Class examination.

Here is the formula:

E▶ Half-power bandwidth $= \dfrac{f_r}{Q}$;　　where: f_r is the resonant frequency in **kHz**
　　　　　　　　　　　　　　　　　　　　　Q is a **quality factor** of the circuit

This is a handy formula to allow you to look at a circuit design and see what frequencies are covered in its half-power bandwidth. Since the answers come out in kHz, you will need to *take their stated resonant frequency in MHz and convert it to kHz by moving the decimal point three places to the right* before you solve the problem. Here's how you do it on your calculator: Clear, *1800 ÷ 95 = 18.94 kHz* so the answer is 18.9 kHz. Yikes, take a look at the possible answers and see if you goofed up on moving the decimal point! **ANSWER A.**

E5A11 What is the half-power bandwidth of a parallel resonant circuit that has a resonant frequency of 7.1 MHz and a Q of 150?

A. 157.8 Hz.　　　　　　　　　　　C. 47.3 kHz.
B. 315.6 Hz.　　　　　　　　　　　D. 23.67 kHz.

7.1 MHz is 7100 kHz, divided by 150. Remember for kHz from MHz, move three decimal places to the right. Clear, *7100 ÷ 150 = 47.33*. **ANSWER C.**

Half-Power Bandwidth

An amplifier's voltage gain will vary with frequency as shown. At the cutoff frequencies, the voltage gain has dropped to 0.707 of what it is in the mid-band. These frequencies, f_1 and f_2, are sometimes called the half-power frequencies.

If the output voltage is 10 volts across a 100-ohm load when the gain is A at the mid-band, then the power output, P_O, at mid-band is:

$$P_O = \frac{E^2}{R} = \frac{10^2}{100} = \frac{100}{100} = 1 \text{ Watt}$$

At the cutoff frequency, the output voltage will be 0.707 of what it is at the mid-band; therefore, 7.07 volts. The power output is:

$$P_{0.707} = \frac{(7.07)^2}{100} = \frac{50}{100} = 0.5 \text{ Watt}$$

The power output at the cutoff frequency points is one-half the mid-band power. The half-power bandwidth is between frequencies f_1 and f_2.

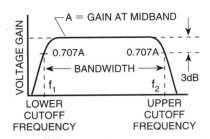

Power Ratio in dB

$$P_{dB} = 10 \log_{10} \frac{P_O}{P_{0.707}}$$

$$= 10 \log_{10} \frac{1}{0.5}$$

$$= 10 \log_{10} 2$$

$$= 10 \times 0.301$$

$$= 3 \text{ db}$$

The power output at the 0.707 frequencies is 3 db down from the mid-band power.

E5A12 What is the half-power bandwidth of a parallel resonant circuit that has a resonant frequency of 3.7 MHz and a Q of 118?

A. 436.6 kHz.   C. 31.4 kHz.

B. 218.3 kHz.   D. 15.7 kHz.

With this many half-power bandwidth questions in the pool, you can count on at least one being on your upcoming Extra Class Element 4 exam. With this question, go from MHz to kHz, and then divide it by the Q. Clear, enter *3700 ÷ 118 = 31.36*. Round the answer to 31.4. **ANSWER C.**

E5A13 What is the half-power bandwidth of a parallel resonant circuit that has a resonant frequency of 14.25 MHz and a Q of 187?

A. 38.1 kHz.   C. 1.332 kHz.

B. 76.2 kHz.   D. 2.665 kHz.

Here is your last half-power bandwidth problem. Hopefully by now you have memorized the formula of E5A10. Remember the conversion to kHz so the general steps are: *MHz to kHz, enter the number in the calculator and divide it by the Q.* There is your answer, *76.2, in kHz*. **ANSWER B.**

E5A14 What is the resonant frequency of a series RLC circuit if R is 22 ohms, L is 50 microhenrys and C is 40 picofarads?

A. 44.72 MHz.   C. 3.56 MHz.

B. 22.36 MHz.   D. 1.78 MHz.

The schematic diagram shown is a simple series resonant circuit. Don't panic. The following resonance formula will carry you through many questions on series and parallel circuits – one of which may be on your exam. You can throw away the resistance found in the parallel or series diagrams. You won't need it for the resonance formula. Here is the resonance formula, which you should memorize:

$$f_r = \frac{10^6}{2\pi\sqrt{LC}}$$

where: f_r is resonant frequency in **kHz**

$2\pi = 6.28$

L is inductance in **microhenrys**

C is capacitance in **picofarads**

Series Resonant Circuit

Your first step is multiplying L × C. For this question, that is 50 × 40 = 2000. Now we need the square root of this, since L and C are under the square root sign. With 2000 showing on your calculator, simply press the square root key. Presto, you should have 44.72. Now multiply by 2π, which is 6.28 × 44.72 = 280.84. This leaves us with 1,000,000 ÷ 280.84. Press your clear button, and enter 1,000,000 ÷ 280.84 = 3560.7463 kHz. Your answer came out in kHz, but they are looking for MHz. No problem here – simply move the decimal point three places to the left, and you get answer C, *3.56MHz*. That's all there is to it! **ANSWER C.**

E5A15 What is the resonant frequency of a series RLC circuit if R is 56 ohms, L is 40 microhenrys and C is 200 picofarads?
A. 3.76 MHz.
C. 11.18 MHz.
B. 1.78 MHz.
D. 22.36 MHz.

Use the formula at E5A14. Solve the denominator first, then divide it into 106. Here are the calculator keystrokes using a calculator that has a square root ($\sqrt{\ }$) key: Clear, 40 × 200 = $\sqrt{\ }$ × 6.28 = 561.7. Remember 561.7. Clear, *1,000,000 ÷ 561.7 = 1780 kHz = 1.78 MHz.* **ANSWER B.**

E5A16 What is the resonant frequency of a parallel RLC circuit if R is 33 ohms, L is 50 microhenrys and C is 10 picofarads?
A. 23.5 MHz.
C. 7.12 kHz.
B. 23.5 kHz.
D. 7.12 MHz.

Use the formula at E5A14. First multiply microhenrys and picofarads, take their square root, multiply by 6.28, and then divide your answer into 1,000,000. Finally, change kHz to MHz. Easy! Here are the keystrokes: Clear, 50 x 10 = $\sqrt{\ }$ × 6.28 = 140.4. Remember 140.4. Clear, *1,000,000 ÷ 140.4 = 7122 kHz which is 7.12 MHz.* C is not correct, it's in kHz. **ANSWER D.**

E5A17 What is the resonant frequency of a parallel RLC circuit if R is 47 ohms, L is 25 microhenrys and C is 10 picofarads?
A. 10.1 MHz.
C. 10.1 kHz.
B. 63.2 MHz.
D. 63.2 kHz.

Use the formula at E5A14. L × C, $\sqrt{\ }$, multiply by 2p, and then divide your answer into 1,000,000. Remember to change kHz to MHz. The keystrokes are: Clear, 25 × 10 = $\sqrt{\ }$ × 6.28 = 99.29. Clear, *1,000,000 ÷ 99.3 = 10070.5 kHz which is 10.1 MHz.* Answer C is not correct. **ANSWER A.**

E5B01 What is the term for the time required for the capacitor in an RC circuit to be charged to 63.2% of the applied voltage?
A. An exponential rate of one.
C. One exponential period.
B. One time constant.
D. A time factor of one.

In an RC circuit, assuming there is no initial charge on the capacitor, *it takes one time constant to charge a capacitor to 63.2 percent of its final supply voltage* value. **ANSWER B.**

E5B02 What is the term for the time it takes for a charged capacitor in an RC circuit to discharge to 36.8% of its initial voltage?
A. One discharge period.
B. An exponential discharge rate of one.
C. A discharge factor of one.
D. One time constant.

Consider the filter capacitors in a power supply, where the capacitors discharge when the input power is removed. The time it takes a capacitor in an RC circuit to *discharge to 36.8 percent of its initial value* of stored charge *is one time constant*. **ANSWER D.**

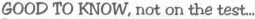

GOOD TO KNOW, not on the test...

To calculate the percentage of the supply voltage after two time constants in an RC circuit, write down the percent after a single time constant as 63.2 percent. The remaining percent of charge that the capacitor must charge is 36.8 percent. It is found by deducting 63.2 percent from 100 percent. Since 36.8 percent is the final value to which the capacitor will charge after it has charged to 63.2 percent, in the next time constant the capacitor will charge to 63.2 percent of the 36.8 percent. In other words, another percent charge is added to the capacitor in the second time constant equal to:

36.8% × 63.2% = 23.26% (rounded to 23.3%)

This really needs to be calculated in decimal format as:

0.368 × 0.632 = 0.2326

Percent values are converted to decimals by moving the decimal point two places to the left. Decimal values are converted to percent values by multiplying by 100 (moving decimal point to right two places.)

Since the final percent of charge in two time constants is desired, it is only necessary to add the two percent charges together:

	Percent	Decimal
Charge in first time constant =	63.2%	0.632
Charge in second time constant =	23.3%	0.233
Total charge after two time constants	86.5%	0.865

The process can be continued for the third time constant. Since the capacitor has charged to 86.5 percent in two time constants, if you start from that point the value of the third charge would be another 13.5 percent. Therefore, in the next time constant, the third, the capacitor would charge another:

13.5% × 63.2% = 8.53% (rounded to 8.5%)

After the third time constant the total charge is:

86.5% + 8.5% = 95%

If the process is continued through five time constants, the capacitor charge is 99.3 percent. Therefore, in electronic calculations, the capacitor is considered to be fully charged (or discharged) after five time constants

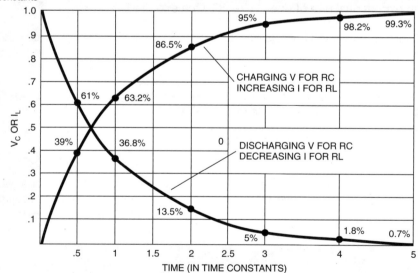

Time Constants Showing V and I Percentages

E5B03 The capacitor in an RC circuit is discharged to what percentage of the starting voltage after two time constants?

A. 86.5%. C. 36.8%.
B. 63.2%. D. 13.5%.

This question asks to what percent the capacitor has *discharged after two time constants*. The capacitor discharges the same percent in a time constant as it charges (for the same RC circuit values of course.) However, the question is asking for the amount of charge remaining. At one time constant, the capacitor discharges 63.2%, therefore the percent charge remaining on the capacitor is 36.8% (100% - 63.2%). Thus, the capacitor on discharge, discharges another 23.3% in the second time constant. After two time constants the capacitor has discharged by 86.5%; therefore, *the remaining charge is 13.5%* (100% - 86.5%). In five time constants the capacitor would have discharged 99.3% of its charge, so 0.7% of charge remains – it is considered to be fully discharged. **ANSWER D.**

E5B04 What is the time constant of a circuit having two 220-microfarad capacitors and two 1-megohm resistors, all in parallel?

A. 55 seconds. C. 440 seconds.
B. 110 seconds. D. 220 seconds.

Here is an RC circuit with all components in parallel. It will be easy if you remember that two equal capacitors in parallel have a total capacitance of twice the value of one of the capacitors, and that two equal value resistors in parallel have a total resistance equal to one-half the value of one of the resistors. And more good news – "megs" are being multiplied by "micros" so the powers of ten after the numerical value cancel. Two 220 microfarad capacitors in parallel is a larger 440 microfarad capacitor. Two one-megohm resistors in parallel is a resistance of 0.5 megohms. *Solve for the time constant using the simple formula $T = R \times C$.* So you multiply 0.5×440 ("megs" times "micros") *and you end up with 220 seconds* as the correct answer. The calculator keystrokes are: Clear, 0.5 × 440 = 220. **ANSWER D.**

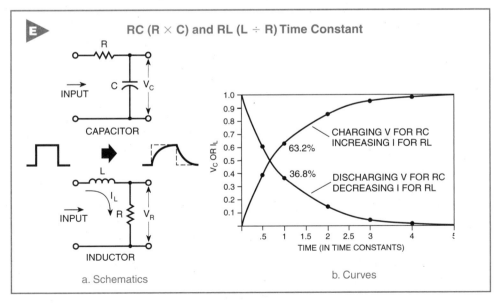

RC (R × C) and RL (L ÷ R) Time Constant

a. Schematics

b. Curves

GOOD TO KNOW, not on the test...

When a resistor is in parallel with a capacitor, the discharging curve from an initial voltage across the capacitor is as shown. The value of a resistor in ohms times a capacitor in farads is called an RC time constant (in seconds). In one RC time constant, the capacitor will discharge 63% of its initial value, leaving a 37% charge. In 5RC, the capacitor is considered totally discharged.

What value of resistor should be used to prevent the capacitor from discharging too much voltage between peaks of the incoming IF signal of 500 kHz? The capacitor has a value of 0.01 microfarads.

Solution:

The time between peaks (the period) of a 500 kHz input signal is

$$\frac{1}{500 \times 10^3} = \frac{1}{5 \times 10^5} = 2 \times 10^{-6}, \text{ or 2 microseconds.}$$

To make the ripple between peaks (the discharge of the capacitor) small, the RC time constant should be 10 times the period of the input waveform. Therefore:

$$RC = 20 \times 10^{-6}, \text{ and since } C = 0.01 \times 10^{-6}$$

$$R \times 0.01 \times 10^{-6} = 20 \times 10^{-6}$$

$$R = \frac{20 \times 10^{-6}}{0.01 \times 10^{-6}} = \frac{20}{1 \times 10^{-2}} = 20 \times 10^2 = 2 \text{ kilohms}$$

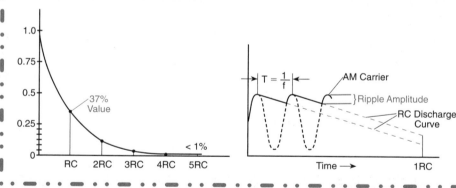

E5B05 How long does it take for an initial charge of 20 V DC to decrease to 7.36 V DC in a 0.01-microfarad capacitor when a 2-megohm resistor is connected across it?

A. 0.02 seconds.

B. 0.04 seconds.

C. 20 seconds.

D. 40 seconds.

Note from the time-constant chart that a capacitor discharges (or charges) 63.2% in one time constant. Thus, after discharging for one time constant, 36.8% of the charge remains on the capacitor. Therefore, in one time constant the 20 volts will drop down to 36.8% or 7.36 volts (0.368 x 20 = 7.36). Write down one time constant = 7.36 volts for later reference. Since 7.36 volts matches the voltage referred to in the question, the time asked for is one time constant. *Use the "megs" times "micros" again to calculate the time constant as 2 × 0.01 = 0.02 seconds.* That's the correct answer. **ANSWER A.**

E5B06 How long does it take for an initial charge of 800 V DC to decrease to 294 V DC in a 450-microfarad capacitor when a 1-megohm resistor is connected across it?

A. 4.50 seconds.

B. 9 seconds.

C. 450 seconds.

D. 900 seconds.

Since we're having so much fun, let's go for some new numbers, such as 800 volts DC and a 450 microfarad capacitor with a 1 megohm resistor connected across it. Let's continue the same steps as in the previous questions. Since the voltage across a capacitor drops to 36.8% of its initial value, what will the voltage be across the capacitor in one time constant? *800 × 0.368 = 294.4 volts*. Write down 294 volts = one time constant. That's the voltage for this question so we know the time wanted is one time constant. What's the time constant? *Using "megs" × "micros", the time constant = 1 × 450 = 450 seconds*. **ANSWER C.**

E5D01 What is the result of skin effect?

A. As frequency increases, RF current flows in a thinner layer of the conductor, closer to the surface.

B. As frequency decreases, RF current flows in a thinner layer of the conductor, closer to the surface.

C. Thermal effects on the surface of the conductor increase the impedance.

D. Thermal effects on the surface of the conductor decrease the impedance.

If you ever had a chance to inspect an old wireless station, you would have seen that the antenna "plumbing" leading out of the transmitter was usually constructed of hollow copper tubing. *Radio frequency current always travels along the thinner outside layer of a conductor*, and as frequency increases, there is almost no current in the center of the conductor. The higher the frequency, the greater the *skin effect*. This is why we use wide copper ground foil to minimize the resistance to AC current that we need to pass to ground. **ANSWER A.**

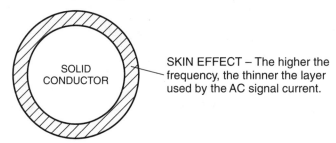

SKIN EFFECT – The higher the frequency, the thinner the layer used by the AC signal current.

SOLID CONDUCTOR

Cross-Section of Wire Conductor Showing Skin Effect

E5D02 Why is the resistance of a conductor different for RF currents than for direct currents?

A. Because the insulation conducts current at high frequencies.

B. Because of the Heisenburg Effect.

C. Because of skin effect.

D. Because conductors are non-linear devices.

The higher the frequency of an *alternating current (AC)* signal, the greater the *skin effect*. For DC, the internal cross-sectional area of a conductor is very important, but for radio frequency signals on the amateur bands, it's not important. **ANSWER C.**

E5D03 What device is used to store electrical energy in an electrostatic field?
    A. A battery.   C. A capacitor.
    B. A transformer.   D. An inductor.

A *capacitor* is made up of parallel plates separated by a dielectric (non-conductor). When a voltage is placed across a capacitor, *energy is stored in the electrostatic field* developed between the capacitor plates. As a reminder, notice the letters "AC" in the word capacitor, and the letters "ATIC" in the word electrostatic. It should lead you to the correct answer. **ANSWER C.**

E5D04 What unit measures electrical energy stored in an electrostatic field?
    A. Coulomb.   C. Watt.
    B. Joule.   D. Volt.

The joule is a measure of energy, or the capacity to do work. The amount of electrical energy stored in an electrostatic field associated with capacitors is expressed in joules. If a *capacitor* is charged with one coulomb of electrons in one second by a voltage of one volt, one joule of energy is stored in the capacitor's electrostatic field. Using energy at a *joule* per second is one watt of electrical power. **ANSWER B.**

E5D05 Which of the following creates a magnetic field?
    A. Potential differences between two points in space.
    B. Electric current.
    C. A charged capacitor.
    D. A battery.

In marine ham radio installations, we always route our power cables well away from the ship's compass. Do you know why? When you pull current through a conductor, the *current produces a magnetic field* as a force around the conductor – the more current; the stronger the field. The force field can affect the reading of a magnetic compass. **ANSWER B.**

E5D07 What determines the strength of a magnetic field around a conductor?
    A. The resistance divided by the current.
    B. The ratio of the current to the resistance.
    C. The diameter of the conductor.
    D. The amount of current.

As *increasing current* passes through a conductor, the *strength of the magnetic field* will continue to build. **ANSWER D.**

E5D08 What type of energy is stored in an electromagnetic or electrostatic field?
    A. Electromechanical energy.   C. Thermodynamic energy.
    B. Potential energy.   D. Kinetic energy.

Energy stored in an *electromagnetic field* in an inductor, or an electrostatic field in a capacitor, is called *potential energy*. **ANSWER B.**

E5D06 In what direction is the magnetic field oriented about a conductor in relation to the direction of electron flow?
- A. In the same direction as the current.
- B. In a direction opposite to the current.
- C. In all directions; omnidirectional.
- D. In a direction determined by the left-hand rule.

If you grab a big straight cable with your left hand, with your thumb pointing in the direction the electrons are flowing in the conductor (opposite from conventional current), your fingers will naturally point in the *direction of the magnetic lines* of force. This is known as the *left-hand rule*. Have you ever been to a junk yard and watched the electromagnet coil magically lift a car off the ground and drop it into a nearby box car? When the coil is energized, it becomes a magnet. As quickly as the energy is cut off, magnetism stops, and the car drops off. **ANSWER D.**

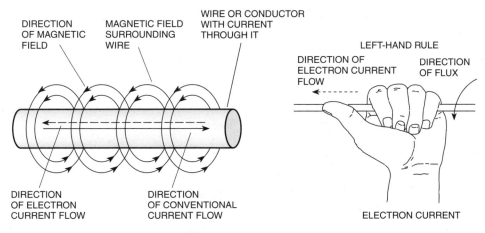

Left-Hand Rule

E5D09 What happens to reactive power in an AC circuit that has both ideal inductors and ideal capacitors?
- A. It is dissipated as heat in the circuit.
- B. It is repeatedly exchanged between the associated magnetic and electric fields, but is not dissipated.
- C. It is dissipated as kinetic energy in the circuit.
- D. It is dissipated in the formation of inductive and capacitive fields.

In a circuit that has both capacitors and inductors, the *reactive power* oscillates back and forth between magnetic and electric fields, and *does not get dissipated*. **ANSWER B.**

E5D10 How can the true power be determined in an AC circuit where the voltage and current are out of phase?
- A. By multiplying the apparent power times the power factor.
- B. By dividing the reactive power by the power factor.
- C. By dividing the apparent power by the power factor.
- D. By multiplying the reactive power times the power factor.

If we *multiply apparent power times the power factor* of an AC voltage and current out of phase, we can *determine true power*. See E5D11 figure. **ANSWER A.**

E5D11 **What is the power factor of an R-L circuit having a 60 degree phase angle between the voltage and the current?**

A. 1.414.
B. 0.866.
C. 0.5.
D. 1.73.

If you are using a scientific calculator, you will find the *cosine of 60 degrees*, given in the problem, *comes out 0.5, answer C*. The power factor is cos ϕ. If you don't have a scientific calculator, you might memorize this table. **ANSWER C.**

ϕ	cos ϕ
30°	0.866
45°	0.707
60°	0.5

Apparent and True Power

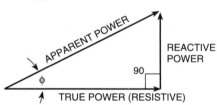

REACTIVE POWER

90

ϕ

TRUE POWER (RESISTIVE)

True power is power dissipated as heat, while reactive power of a circuit is stored in inductances or capacitances and then returned to the circuit.

Apparent power is voltage times current without taking into account the phase angle between them. According to right angle trigonometry,

$$\text{cosine } \phi = \frac{\text{side adjacent}}{\text{hypotenuse}}$$

therefore,

$$\text{cosine } \phi = \frac{\text{True Power}}{\text{Apparent Power}}$$

therefore,

True Power = Apparent Power × Cosine ϕ

Cosine ϕ is called the *power factor* (PF) of a circuit.

Since apparent power is E × I,

True Power = E × I × Cos ϕ

or

True Power = E × I × PF

E5D12 **How many watts are consumed in a circuit having a power factor of 0.2 if the input is 100-V AC at 4 amperes?**

A. 400 watts.
B. 80 watts.
C. 2000 watts.
D. 50 watts.

This problem is based on the formula:

 True power = E (volts) × I (amps) × PF (power factor).

True power works out to be *100 volts × 4 amps × 0.2, giving us 80 watts*. Easy enough, right? **ANSWER B.**

E5D13 How much power is consumed in a circuit consisting of a 100 ohm resistor in series with a 100 ohm inductive reactance drawing 1 ampere?

    A. 70.7 Watts.   C. 141.4 Watts.

    B. 100 Watts.   D. 200 Watts.

The coil, with 100 ohms inductive reactance, will not factor in this problem, as a pure inductance will not consume power in this circuit. Ohm's Law, $P = I^2 \times R$, will result in $1 \times 1 = 1$, times R of 100 ohms resistance, equal to *100 watts* of power consumed. **ANSWER B.**

$$P = I^2 \times R \qquad \text{Where: } P = \text{Power}$$
$$P = (1 \times 1) \times 100 \qquad I = \text{Amps}$$
$$P = 1 \times 100 \qquad R = \text{Resistance in ohms}$$
$$P = 100 \text{ watts}$$

E5D14 What is reactive power?

    A. Wattless, nonproductive power.

    B. Power consumed in wire resistance in an inductor.

    C. Power lost because of capacitor leakage.

    D. Power consumed in circuit Q.

The letters AC are found in the word *reactive* (meaning out-of-phase). It indicates *non-productive power* produced in circuits containing inductors and capacitors. Since out-of-phase power in inductors will tend to cancel out-of-phase power in capacitors, reactive power is *wattless*. It is not converted to heat and dissipated. It is energy being stored temporarily in a field and then returned to the circuit. It circulates back and forth in a coil's magnetic field and/or a capacitor's electrostatic field. **ANSWER A.**

E5D15 What is the power factor of an RL circuit having a 45 degree phase angle between the voltage and the current?

    A. 0.866.   C. 0.5.

    B. 1.0.   D. 0.707.

Using your scientific calculator, or from memory, remember that *the cosine of 45 degrees is 0.707.* Think of an airplane – the very popular 707 jet aircraft – taking off at an *angle of 45 degrees.* **ANSWER D.**

E5D16 What is the power factor of an RL circuit having a 30 degree phase angle between the voltage and the current?

    A. 1.73.   C. 0.866.

    B. 0.5.   D. 0.577.

At *30 degrees*, the *power factor is 0.866.* If you don't pass your Extra Class exam, you could be 86'd out of the room! **ANSWER C.**

E5D17 How many watts are consumed in a circuit having a power factor of 0.6 if the input is 200V AC at 5 amperes?

    A. 200 watts.   C. 1600 watts.

    B. 1000 watts.   D. 600 watts.

Here are 2 simple problems that are based on the formula:

     True power = E (volts) $\times$ I (amps) $\times$ PF (power factor).

Therefore, *200 volts $\times$ 5 amps $\times$ 0.6 = 600 watts.* Easy, huh? **ANSWER D.**

E5D18 How many watts are consumed in a circuit having a power factor of 0.71 if the apparent power is 500 VA?

A. 704 W.

B. 355 W.

C. 252 W.

D. 1.42 mW.

Here they have already calculated apparent power at 500 VA (Volt Amps). *Multiply apparent power by power factor 0.71*, and you end up with *355 watts* consumed in the circuit. **ANSWER B.**

E4E04 How can conducted and radiated noise caused by an automobile alternator be suppressed?

A. By installing filter capacitors in series with the DC power lead and by installing a blocking capacitor in the field lead.

B. By installing a noise suppression resistor and a blocking capacitor in both leads.

C. By installing a high-pass filter in series with the radio's power lead and a low-pass filter in parallel with the field lead.

D. By connecting the radio's power leads directly to the battery and by installing coaxial capacitors in line with the alternator leads.

Wire your radio's red and black leads directly to your battery positive and negative terminals using the shortest possible run. Be sure you put a fuse right next to the battery positive pick-off point for safety. Also, it is recommended that you *install coaxial capacitors in the alternator leads*. **ANSWER D.**

E4E05 How can noise from an electric motor be suppressed?

A. By installing a high pass filter in series with the motor's power leads.

B. By installing a brute-force AC-line filter in series with the motor leads.

C. By installing a bypass capacitor in series with the motor leads.

D. By using a ground-fault current interrupter in the circuit used to power the motor.

One way of reducing noise coming up the line from an electric motor is to install an AC-line filter in series with the motor leads. The incorrect answers have got the right components, but the wrong usage. We use RF beads to keep RF off of specific circuits, and we use bypass capacitors in parallel to bypass noise. As for the ground-fault interrupter, this won't help with motor noise. Go with the *brute-force AC-line filter in series*. **ANSWER B.**

E6D06 What core material property determines the inductance of a toroidal inductor?

A. Thermal impedance.

B. Resistance.

C. Reactivity.

D. Permeability.

The key words in the question are "*core material property*," and the material that determines the *inductance of a toroid is the permeability* of the iron donut. **ANSWER D.**

Ferrite Toroids
Courtesy of Palomar Engineers

E6D07 What is the usable frequency range of inductors that use toroidal cores, assuming a correct selection of core material for the frequency being used?
  A. From a few kHz to no more than 30 MHz.
  B. From less than 20 Hz to approximately 300 MHz.
  C. From approximately 10 Hz to no more than 3000 kHz.
  D. From about 100 kHz to at least 1000 GHz.
If you carefully select the core material, the *frequency range* of the inductor will span approximately *20 Hz to 300 MHz*. Refer to the charts of core size and core specs when using popular toroidal cores. **ANSWER B.**

E6D08 What is one important reason for using powdered-iron toroids rather than ferrite toroids in an inductor?
  A. Powdered-iron toroids generally have greater initial permeabilities.
  B. Powdered-iron toroids generally maintain their characteristics at higher currents.
  C. Powdered-iron toroids generally require fewer turns to produce a given inductance value.
  D. Powdered-iron toroids have higher power handling capacity.
When designing high current circuits, such as those in linear amplifiers, *powdered-iron toroids offer the greatest high current capabilities* while maintaining a stable inductance. Powdered-iron toroids also offer good temperature stability.
ANSWER B.

E6D16 What is one reason for using ferrite toroids rather than powdered-iron toroids in an inductor?
  A. Ferrite toroids generally have lower initial permeabilities.
  B. Ferrite toroids generally have better temperature stability.
  C. Ferrite toroids generally require fewer turns to produce a given inductance value.
  D. Ferrite toroids are easier to use with surface mount technology.
Another sound reason to use ferrite toroids rather than powdered-iron toroids is the fact that *ferrite toroids require less turns of wire to produce a given inductance* value. **ANSWER C.**

E6D10 What is a primary advantage of using a toroidal core instead of a solenoidal core in an inductor?
  A. Toroidal cores confine most of the magnetic field within the core material.
  B. Toroidal cores make it easier to couple the magnetic energy into other components.
  C. Toroidal cores exhibit greater hysteresis.
  D. Toroidal cores have lower Q characteristics.
The *toroidal inductor concentrates the magnetic field within the core material*, protecting other nearby components from stray magnetic fields from that inductor.
ANSWER A.

E6D11 How many turns will be required to produce a 1-mH inductor using a ferrite toroidal core that has an inductance index (AL) value of 523 millihenrys/1000 turns?

A. 2 turns.
B. 4 turns.
C. 43 turns.
D. 229 turns.

Here is the formula to calculate the number of turns required to produce a specific inductance where you know the inductance index, AL, of the material. AL is determined and published by the core manufacturer and is an index that is the microhenrys of inductance per 1000 turns or millihenrys of inductance per 1000 turns of single layer wire on the core.

$$N = 1000 \sqrt{\frac{L}{A_L}}$$

where: N is number of turns needed
L is inductance in **millihenrys** of core inductor
A_L (A sub L) is inductance index in **mH per 1000 turns**

$$N = 1000 \sqrt{\frac{1}{523}}$$

$$N = 1000 \sqrt{0.001912} = 1000 \sqrt{19.12 \times 10^{-2}}$$

$$N = 1 \times 10^3 \times 4.37 \times 10^{-2}$$

$$N = 4.37 \times 10^1 = 43.7$$

Here's how to work it out on your calculator. Got a calculator, right? Hit the clear button several times. Then enter *1 ÷ 523 =*. The result is *0.001912. Press* $\sqrt{}$ and the result is *0.0437264. Press* × *1000 = N for the answer 43.7264*. Your closest correct answer is 43. **ANSWER C.**

E6D12 How many turns will be required to produce a 5-microhenry inductor using a powdered-iron toroidal core that has an inductance index (AL) value of 40 microhenrys/100 turns?

A. 35 turns.
B. 13 turns.
C. 79 turns.
D. 141 turns.

This time we are talking about calculating the number of turns in a powdered-iron toroidal core. What we just worked with on in the last question as a ferrite toroidal core. The tip-off if you miss the wording "powdered-iron" is that we only have 5 microhenrys after a bunch of turns, rather than millihenrys with a ferrite core. Powdered-iron has better temperature stability, but the overall inductance will be dramatically less. Here's your next formula:

$$N = 1000 \sqrt{\frac{L}{A_L}}$$

where: N is number of turns
L is inductance in **microhenrys**
A_L (A sub L) is inductance index in **µH per 100 turns**

$$N = 1000 \sqrt{\frac{5}{40}}$$

$$N = 100 \sqrt{0.125} = 100 \sqrt{12.5 \times 10^{-2}}$$

$$N = 3.53 \times 10^{-1} \times 10^2 = 3.53 \times 10^1 = 35.3$$

Now that wasn't so hard, was it? The formulas are almost the same, and luckily on both of these problems if you should reverse the formulas in error, they do not have an incorrect answer just waiting for you to get it wrong. **ANSWER A.**

Maximum Number of Turns Using Popular Toroidal Cores

Wire Size	T-80	T-68	T-50	T-37	T-25
12	14	9	6	3	
14	18	13	8	5	1
16	24	17	13	7	2
18	32	23	18	10	4
22	53	38	30	19	9
28	108	80	64	42	23
32	171	127	103	68	38

A_Λ — µH per 100 Turns — for Popular Toroidal Cores

CORE SPECS ▶	MIX 6	MIX 10	MIX 12	MIX 0			
Core Size ▼	*$\mu = 8.5$	$\mu = 6.0$	$\mu = 4.0$	$\mu = 1.0$	SIZE (INCHES)		
	2-30 MHz	10-100 MHz	20-200 MHz	50-250 MHz	OD	ID	Ht
T-80	45	32	22	8.5	0.80	0.50	0.25
T-68	47	32	21	7.5	0.68	0.37	0.19
T-50	40	31	18	6.4	0.50	0.30	0.19
T-37	30	25	15	4.9	0.37	0.20	0.13
T-25	27	19	12	4.5	0.25	0.12	0.10

*permeability

Components in Your New Rig

FREE ELECTRONS!
JUST THINK...
FREE!!!

E6A01 In what application is gallium arsenide used as a semiconductor material in preference to germanium or silicon?

    A. In high-current rectifier circuits.
    B. In high-power audio circuits.
    C. At microwave frequencies.
    D. At very low frequency RF circuits.

Here's a new word for your amateur radio vocabulary: GaAsFET. The GaAs stands for *gallium arsenide*, and FET stands for field effect transistor. We use the *GaAsFET* transistor in VHF, UHF, and *microwave receivers*. This transistor offers excellent signal gain with an extremely-low noise figure. **ANSWER C.**

E6A02 Which of the following semiconductor materials contains excess free electrons?

    A. N-type.   C. Bipolar.
    B. P-type.   D. Insulated gate.

If there are *more free electrons* orbiting the nucleus, it is an N-type material. **ANSWER A.**

You will have one question from this group about semiconductor materials. Remember this phrase, "ANFREE" for an N-type material where antimony atoms or arsenic atoms have been added, and there are more free electrons in an N-type semiconductor. "PIUMH" (pronounced like sneezing) is a P-type material produced by adding either gallIUM or indIUM, producing more "Potholes" than free electrons. Also, anything that PROVIDES is a donor, and an Acceptor impurity will Add holes. Got it? Good reading about semiconductors is found in the latest edition of the American Radio Relay League Handbook for Radio Amateurs.

Bipolar Transistor Basics – PNP and NPN

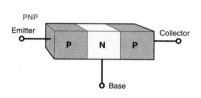

a. Junction Structure

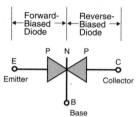

b. Diode Juction Equivalent

c. PNP Symbol

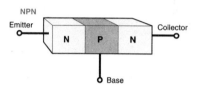

d. Junction Structure

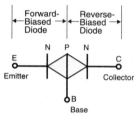

e. Diode Junction Equivalent

f. NPN Symbol

There are two types of transistors, bipolar and field-effect. A bipolar transistor is a combination of two junctions of semiconductor material built into a semiconductor chip (usually silicon). There are two types – PNP and NPN. Their junction structures are shown in Figure a and d, respectively. For a transistor that produces gain, the emitter-base junction is a forward-biased diode, and the collector-base junction is a reverse-biased diode. The diode equivalents of PNP and NPN transistors are shown in Figure b and e, respectively. However, unlike a reverse-biased diode that does not conduct current (except for a small leakage current), the reverse-biased collector-base junction of a bipolar transistor conducts collector current that is controlled by the current into the base at the base-emitter junction. The normal active device operation is shown below in Figure g and h.

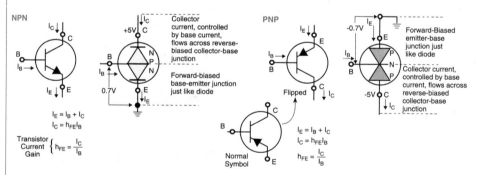

g. NPN Operation

h. PNP Operation

An NPN silicon transistor P base is 0.7 V more positive that its N emitter, and the N collector is several more volts more positive than the emitter. The emitter current is the sum of the base and collector current. The current gain under any DC operating condition is hFE, the ratio of I_C to I_B, and current gains of 50 to 200 are common in modern-day silicon transistors. hFE is actually called "the common-emitter" current gain because the emitter is common in the circuit.

A PNP silicon transistor N base is 0.7 V negative with respect to its emitter in order to have the P emitter more positive than the N base. The P collector is several volts negative from the emitter to keep the collector-base junction reverse biased. As shown in Figure h, the same current equations apply, and the current gain, hFE, is the same. The major difference is in the polarity of the voltages for operation. For NPN common-emitter operation the base and collector voltages are positive with respect to the emitter; while for the PNP the voltages are negative.

Source: *Basic Communications Electronics*, Hudson & Luecke, © 1999 Master Publishing, Inc., Niles, IL

E6A03 What are the majority charge carriers in P-type semiconductor material?

A. Free neutrons.

B. Free protons.

C. Holes.

D. Free electrons.

The charge carriers are in abundance in *P-type* semiconductor materials, the *holes*. **ANSWER C.**

E6A04 What is the name given to an impurity atom that adds holes to a semiconductor crystal structure?

A. Insulator impurity.

B. N-type impurity.

C. Acceptor impurity.

D. Donor impurity.

The *impurity atom that adds holes* to a semiconductor crystal structure is called an *acceptor impurity*. **ANSWER C.**

E6A05 What is the alpha of a bipolar junction transistor?

A. The change of collector current with respect to base current.

B. The change of base current with respect to collector current.

C. The change of collector current with respect to emitter current.

D. The change of collector current with respect to gate current.

In *bipolar* transistors, the term "*alpha*" is the *variation of collector current with respect to emitter current* (CCEC). This is an important consideration when designing a circuit that will utilize a bipolar transistor. **ANSWER C.**

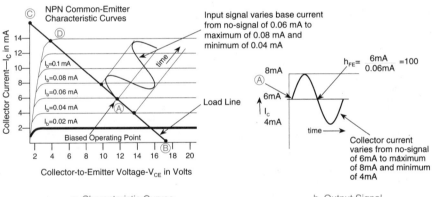

a. Characteristic Curves   b. Output Signal

Biasing sets "No-Signal" operating point

E6A06 What is the beta of a bipolar junction transistor?

A. The frequency at which the current gain is reduced to 1.

B. The change in collector current with respect to base current.

C. The breakdown voltage of the base to collector junction.

D. The switching speed of the transistor.

The bipolar junction transistor could be NPN type, or PNP type, and it is current-operated. The term "*beta*" refers to the *change in current gain and is stated as* a ratio of the *change in collector current to the applied change in base current*. **ANSWER B.**

E6A07 In Figure E6-1, what is the schematic symbol for a PNP transistor?

A. 1. C. 4.
B. 2. D. 5.

Remember this from your Technician Class examination? Which way is the arrow pointing? If it's *pointing in, it is PNP – "P in."* **ANSWER A.**

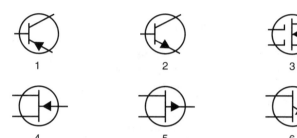

Figure E6-1

E4A10 Which of the following tests establishes that a silicon NPN junction transistor is biased on?

A. Measure base-to-emitter resistance with an ohmmeter; it should be approximately 6 to 7 ohms.
B. Measure base-to-emitter resistance with an ohmmeter; it should be approximately 0.6 to 0.7 ohms.
C. Measure base-to-emitter voltage with a voltmeter; it should be approximately 6 to 7 volts.
D. Measure base-to-emitter voltage with a voltmeter; it should be approximately 0.6 to 0.7 volts.

When an NPN transistor is biased "on," the voltage on the base is positive with respect to the emitter. Free electrons can flow from the emitter to the base, and on the collector, assuming the base to collector is reversed bias. A characteristic of a *forward-biased PN junction* is that the *voltage drop* across it is always approximately *0.6 to 0.7V*. You don't need to remove the transistor from the circuit for this easy transistor check. **ANSWER D.**

E6A08 What term indicates the frequency at which the grounded-base current gain of a transistor has decreased to 0.7 of the gain obtainable at 1 kHz?

A. Corner frequency. C. Beta cutoff frequency.
B. Alpha rejection frequency. D. Alpha cutoff frequency.

Grounded-base is the same as a common base configuration. Common base means alpha. Since *1 kHz* is mentioned in this question, they are referring to the *alpha cutoff frequency* in a transistor. **ANSWER D.**

E6A09 What is a depletion-mode FET?

 A. An FET that exhibits a current flow between source and drain when no gate voltage is applied.

 B. An FET that has no current flow between source and drain when no gate voltage is applied.

 C. Any FET without a channel.

 D. Any FET for which holes are the majority carriers.

Your handheld battery may become depleted if *depletion-mode FETs* are used in many of the circuits. Even though there is no gate voltage applied, *current still continues to flow*. **ANSWER A.**

E6A10 In Figure E6-2, what is the schematic symbol for an N-channel dual-gate MOSFET?

A. 2.	C. 5.
B. 4.	D. 6.

Note that the *arrow is pointing iN*, and there are now *two gates*. **ANSWER B.**

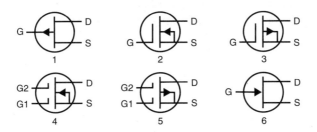

Figure E6-2

E6A11 In Figure E6-2, what is the schematic symbol for a P-channel junction FET?

A. 1.	C. 3.
B. 2.	D. 6.

In the *P-channel junction FET*, the *arrow is Pointing out*. For an N-channel junction FET, the arrow is always pointing iN. **ANSWER A.**

E6A12 Why do many MOSFET devices have internally connected Zener diodes on the gates?

 A. To provide a voltage reference for the correct amount of reverse-bias gate voltage.

 B. To protect the substrate from excessive voltages.

 C. To keep the gate voltage within specifications and prevent the device from overheating.

 D. To reduce the chance of the gate insulation being punctured by static discharges or excessive voltages.

The "front end" transistors of the modern ham transceiver must sometimes sustain major amounts of incoming signals from nearby transmitters, or maybe even a nearby lightning strike. Lightning *static* could destroy a MOSFET if it weren't for the built-in, *gate-protective Zener diodes*. **ANSWER D.**

E6A13 What do the initials CMOS stand for?
A. Common Mode Oscillating System.
B. Complementary Mica-Oxide Silicon.
C. Complementary Metal-Oxide Semiconductor.
D. Common Mode Organic Silicon.

When we talk about transistorized devices, they are in fact semi-conductors. A *CMOS* is made from layers of *metal-oxide semiconductor* material. **ANSWER C.**

E6A14 How does DC input impedance at the gate of a field-effect transistor compare with the DC input impedance of a bipolar transistor?
A. They are both low impedance.
B. An FET has low input impedance; a bipolar transistor has high input impedance.
C. An FET has high input impedance; a bipolar transistor has low input impedance.
D. They are both high impedance.

An *FET* is much easier to work with in circuits than the simple bipolar transistor. Its *higher input impedance* is less likely to load down a circuit that it is attached to, and the FET can also handle big swings in signal levels. **ANSWER C.**

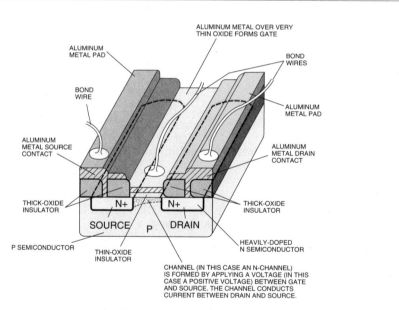

Pictorial of FET Construction (N-Channel Enhancement)

Unlike bipolar transistors that depend on current into the base to control collector current, field-effect transistor current between source and drain is controlled by a voltage on a gate. The basic structure of an N-channel MOSFET is shown here. Heavily-doped N semiconductor material forms source and drain regions in a P semiconductor material substrate. The region between the source and drain is the gate region, where a thin layer of oxide insulates the P semiconductor substrate underneath from a metal plate that is deposited over the thin oxide. A thick oxide layer over the source and drain regions insulates metal connection pads from the substrate. Holes in this thick oxide layer allow the metal pads to contact the source and drain. There are four common types of MOSFETs: P-channel depletion and enhancement mode devices; and N-channel depletion and enhancement mode devices.

Source: *Basic Communications Electronics*, Hudson & Luecke, © 1999 Master Publishing, Inc., Niles, IL

E6A16 What are the majority charge carriers in N-type semiconductor material?

A. Holes.
B. Free electrons.
C. Free protons.
D. Free neutrons.

In an *N-type* semiconductor material, there are more *free electrons* than holes. **ANSWER B.**

E6A15 Which of the following semiconductor materials contains an excess of holes in the outer shell of electrons?

A. N-type.
B. P-type.
C. Superconductor-type.
D. Bipolar-type.

If there are *more holes* than electrons, it is a *P-type* semiconductor. **ANSWER B.**

E6A17 What are the names of the three terminals of a field-effect transistor?

A. Gate 1, gate 2, drain.
B. Emitter, base, collector.
C. Emitter, base 1, base 2.
D. Gate, drain, source.

Gosh darn semiconductors! *Gate, drain, source.* Gosh darn field effect transistor, a semiconductor! **ANSWER D.**

E6B01 What is the most useful characteristic of a Zener diode?

A. A constant current drop under conditions of varying voltage.
B. A constant voltage drop under conditions of varying current.
C. A negative resistance region.
D. An internal capacitance that varies with the applied voltage.

Always remember that a *Zener diode* is for voltage regulation, for *constant voltage* even though current changes. **ANSWER B.**

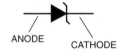

ANODE   CATHODE

Here is the schematic symbol of a Zener diode. Since a diode only passes energy in one direction, look for that one-way arrow, plus a "Z" indicating it is a Zener diode. Doesn't that vertical line look like a tiny "Z".

Zener Diode

E6B02 What is an important characteristic of a Schottky diode as compared to an ordinary silicon diode when used as a power supply rectifier?

A. Much higher reverse voltage breakdown.
B. Controlled reverse avalanche voltage.
C. Enhanced carrier retention time.
D. Less forward voltage drop.

The Schottky diode is regularly found in solar power and power supply circuits because of its lower forward resistance and a slightly higher peak inverse voltage (PIV) rating. They are slightly more expensive than traditional power supply diodes, but I use them regularly in my mobile and home solar system because the *Schottky diode offers minimal voltage drop* while charging my sealed battery system, and prevents current from flowing out of my battery system and back into the solar panel at night. **ANSWER D.**

E6B03 What special type of diode is capable of both amplification and oscillation?
A. Point contact.   C. Tunnel.
B. Zener.   D. Junction.
The *tunnel diode* is commonly used in both *amplifier and oscillator* circuits. The negative resistance region is particularly useful in oscillators. **ANSWER C.**

E6B04 What type of semiconductor device is designed for use as a voltage-controlled capacitor?
A. Varactor diode.   C. Silicon-controlled rectifier.
B. Tunnel diode.   D. Zener diode.
We use the *varactor diode* to tune VHF and UHF circuits by *varying the voltage* applied to the varactor diode. **ANSWER A.**

E6B10 In Figure E6-3, what is the schematic symbol for a light-emitting diode?
A. 1.   C. 6.
B. 5.   D. 7.
The light-emitting diode (LED) is a giveaway on the schematic diagram! Look at the *two arrows indicating an illuminating device.* **ANSWER B.**

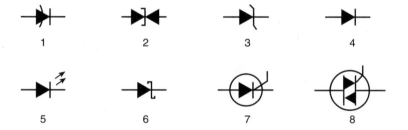

Figure E6-3

E6B06 Which of the following is a common use of a hot-carrier diode?
A. As balanced mixers in FM generation.
B. As a variable capacitance in an automatic frequency control circuit.
C. As a constant voltage reference in a power supply.
D. As a VHF / UHF mixer or detector.
When you get involved with *VHF and UHF* equipment, you'll find *hot-carrier diodes* commonly used as *mixers or detectors* because of their low noise figure characteristics. **ANSWER D.**

E6B07 What is the failure mechanism when a junction diode fails due to excessive current?
A. Excessive inverse voltage.
B. Excessive junction temperature.
C. Insufficient forward voltage.
D. Charge carrier depletion.
Guess what will kill most electronic components? High temperature! The limit of the maximum forward current in a junction diode is the junction temperature. This is why you'll find these diodes mounted to the chassis of your equipment. The

chassis acts as a heat sink to keep the *junction temperature* from *exceeding its maximum limit.* **ANSWER B.**

E6B08 Which of the following describes a type of semiconductor diode?
    A. Metal-semiconductor junction.
    B. Electrolytic rectifier.
    C. CMOS-field effect.
    D. Thermionic emission diode .
The two main categories of *semiconductor diodes* are PN junction, and *metal-semiconductor junction*. The metal semiconductor diode might be found in transistors, power supply rectifier diodes, and photovoltaic panels. **ANSWER A.**

E6B09 What is a common use for point contact diodes?
    A. As a constant current source.
    B. As a constant voltage source.
    C. As an RF detector.
    D. As a high voltage rectifier.
The little *point contact diode*, since it only conducts in one direction, may be used as an *RF detector* in some VHF and UHF equipment. **ANSWER C.**

E6B12 What is one common use for PIN diodes?
    A. As a constant current source.
    B. As a constant voltage source.
    C. As an RF switch.
    D. As a high voltage rectifier.
PIN diodes may be used in small, handheld transceivers for *RF switching*. This gets away from mechanical relays for switching. **ANSWER C.**

E6B05 What characteristic of a PIN diode makes it useful as an RF switch or attenuator?
    A. Extremely high reverse breakdown voltage.
    B. Ability to dissipate large amounts of power.
    C. Reverse bias controls its forward voltage drop.
    D. A large region of intrinsic material.
The PIN diode operating at forward bias will act as a closed switch, passing microwave radio frequency signals. If you don't have transmit/receive relays in that new rig, chances are the silent T/R circuitry is a PIN diode. The *PIN diode* has a *layer of intrinsic, un-doped semi conductor material* between the P and N type materials. Forward bias can turn them into a completed circuit switch, yet a decrease of forward bias may turn it into a variable resistance to RF signals passing through. **ANSWER D.**

E6B11 What is used to control the attenuation of RF signals by a PIN diode?
    A. Forward DC bias current.
    B. A sub-harmonic pump signal.
    C. Reverse voltage larger than the RF signal.
    D. Capacitance of an RF coupling capacitor.
The *PIN diode* can not only work as an RF switch, but also as an *RF attenuator* if the *forward DC bias current* is reduced. With just the right amount of forward bias, the PIN diode may now act as a variable resistance to RF signals. **ANSWER A.**

E6F07 What is a solid state relay?
A. A relay using transistors to drive the relay coil.
B. A device that uses semiconductor devices to implement the functions of an electromechanical relay.
C. A mechanical relay that latches in the on or off state each time it is pulsed.
D. A passive delay line.

The *PIN diode* is sometimes called a *solid state relay*, capable of passing microwave and RF signals when forward biased. The PIN diode is NOT capable of replacing the high voltage and high current handling properties of mechanical relays in automatic antenna tuners. Electromechanical relays in tuners will stick around for some time to come. **ANSWER B.**

E6B13 What type of bias is required for an LED to emit light?
A. Reverse bias.   C. Zero bias.
B. Forward bias.   D. Inductive bias.

The *LED* requires *forward bias* in order to *illuminate*. As indicated in symbol 5 of Figure E6-4, the LED is a diode, and it must be forward biased in order to make it give off light. **ANSWER B.**

E6C01 What is the recommended power supply voltage for TTL series integrated circuits?
A. 12 volts.   C. 5 volts.
B. 1.5 volts.   D. 13.6 volts.

If you ever worked on *TTL (transistor-transistor logic) circuits*, you know the importance of those *5-volt* regulators that always seem to burn up at the wrong time. TTL circuits run on 5.0 volts. **ANSWER C.**

E6C02 What logic state do the inputs of a TTL device assume if they are left open?
A. A logic-high state.
B. A logic-low state.
C. The device becomes randomized and will not provide consistent high or low-logic states.
D. Open inputs on a TTL device are ignored.

Here are several key words – *inputs open and high state*. Remember four letters make up the word open, and four letters make up the word high. **ANSWER A.**

E6C05 Which of the following is an advantage of CMOS logic devices over TTL devices?
A. Differential output capability.
B. Lower distortion.
C. Immune to damage from static discharge.
D. Lower power consumption.

The Complimentary Metal Oxide Semiconductor (CMOS) contains both N-channel and P-channel MOSFETs within a single chip. The greatest advantage of the *CMOS* is its *extremely low power consumption* in that new handheld you just bought for the kids. **ANSWER D.**

E6C06 Why do CMOS digital integrated circuits have high immunity to noise on the input signal or power supply?
A. Larger bypass capacitors are used in CMOS circuit design.
B. The input switching threshold is about two times the power supply voltage.
C. The input switching threshold is about one-half the power supply voltage.
D. Input signals are stronger.

The *CMOS* device is relatively *immune to noise* because its *switching threshold is one-half the power supply voltage*. As a result, power supply noise transients or input transients much larger than other logic circuit types will not cause a transition in the input state. **ANSWER C.**

E6E04 What is the most common input and output impedance of circuits that use MMICs?
A. 50 ohms.
B. 300 ohms.
C. 450 ohms.
D. 10 ohms.

The beauty of the *MMIC* (Monolithic Microwave Integrated Circuit) is that it has a *characteristic impedance of 50 ohms*, and this what the microwavers love!.
ANSWER A.

E6E11 Which of the following materials is likely to provide the highest frequency of operation when used in MMICs?
A. Silicon.
B. Silicon nitride.
C. Silicon dioxide.
D. Gallium nitride.

Gallium nitride is the latest crystal technology for high-temperature and high-voltage applications in both microwave amplifiers as well as in your laser disk player! GaN crystal technology also is found in solar panels up in outer space, where the output remains steady while the panels get toasted by the Sun. LED high-intensity optical shaft encoder circuits may also use the *galium nitride* LED. For ham use, we find this crystal in the tiny, pill-sized *MMIC* packages. **ANSWER D.**

E6E07 Which of the following techniques is typically used to construct a MMIC-based microwave amplifier?
A. Ground-plane construction.
B. Microstrip construction.
C. Point-to-point construction.
D. Wave-soldering construction.

Up at microwave frequencies, we use *microstrip* construction, which is ideal for the MMIC. **ANSWER B.**

E6E08 How is power-supply voltage normally furnished to the most common type of monolithic microwave integrated circuit (MMIC)?
A. Through a resistor and/or RF choke connected to the amplifier output lead.
B. MMICs require no operating bias.
C. Through a capacitor and RF choke connected to the amplifier input lead.
D. Directly to the bias-voltage (VCC IN) lead.

The operating bias voltage for the MMIC is developed through a *resistor and RF choke connected to the amplifier output lead*. **ANSWER A.**

HERE IS A HELPFUL WEBSITE RELATED TO COMPONENTS IN YOUR NEW RIG
http://en.wikipedia.org/wiki/input_offset/voltage

Logically Speaking of Counters

MY FLIP-FLOP'S
TRUTH TABLE TELLS
ME YOU'RE GOING TO
ACE THE EXAM!

E6C07 In Figure E6-5, what is the schematic symbol for an AND gate?
    A. 1.   C. 3.
    B. 2.   D. 4.
The *AND* gate looks like the *letter D*, with two or more inputs, and one output.
ANSWER A.

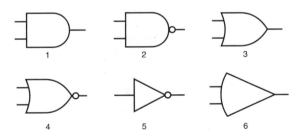

Figure E6-5

E6C08 In Figure E6-5, what is the schematic symbol for a NAND gate?
    A. 1.   C. 3.
    B. 2.   D. 4.
Since the *NAND* gate has a *logic "0"* at its output, spot the little *tiny "o" on its nose*.
It looks like the letter D, the same as the AND gate, except for the small circle at its
output. **ANSWER B.**

E6C09 In Figure E6-5, what is the schematic symbol for an OR gate?

A. 2.
B. 3.
C. 4.
D. 6.

The word "OR" starts with an "O", and its input is curved like a portion of the letter "O". The *OR gate* symbol has a *curved input* and a *pointed output.* **ANSWER B.**

E6C10 In Figure E6-5, what is the schematic symbol for a NOR gate?

A. 1.
B. 2.
C. 3.
D. 4.

Since the *NOR gate* will give us a logic "0" at its output if any inputs are logic "1," look for the little "o" on the nose, and a concave input. It *looks just like an OR gate*, except for the *small circle on its nose.* **ANSWER D.**

E6C11 In Figure E6-5, what is the schematic symbol for the NOT operation (inverter)?

A. 2.
B. 4.
C. 5.
D. 6.

The *NOT gate is triangular* in shape and has a little "o" on its nose. It looks altogether different from any of the other logic gates. You'll probably have one question based on gate symbols. Make up flashcards, and see if you can easily call out which is which. **ANSWER C.**

E7A01 Which of the following is a bistable circuit?

A. An "AND" gate.
B. An "OR" gate.
C. A flip-flop.
D. A clock.

A bistable circuit is a *flip-flop*. Bistable stands for *two stable states.* **ANSWER C.**

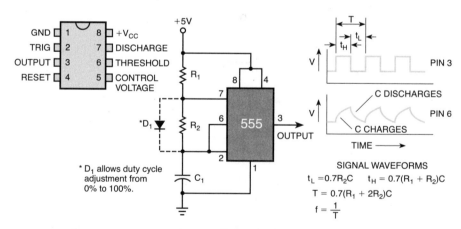

An IC 555 timer can be used as a bistable, astable, or monostable multivibrator. This schematic shows it wired as an astable multivibrator. The resulting astable voltage waveforms at pins 3 and 6 are shown.

Source: *Basic Digital Electronics*, Evans, ©1996, Master Publishing, Inc., Niles, IL

E7A02 How many output level changes are obtained for every two trigger pulses applied to the input of a T flip-flop circuit?

A. None.
B. One.

C. Two.
D. Four.

Two trigger pulses, two output level changes. The T FF toggles from one state to another for each input pulse. **ANSWER C.**

E7A03 Which of the following can divide the frequency of a pulse train by 2?

A. An XOR gate.
B. A flip-flop.

C. An OR gate.
D. A multiplexer.

A *flip-flop* can be used to *divide a pulse train* (the frequency of an AC signal). **ANSWER B.**

E7A04 How many flip-flops are required to divide a signal frequency by 4?

A. 1.
B. 2.

C. 4.
D. 8.

Since a flip-flop has two stable states, you will need *two flip-flops to divide a signal frequency by four*. **ANSWER B.**

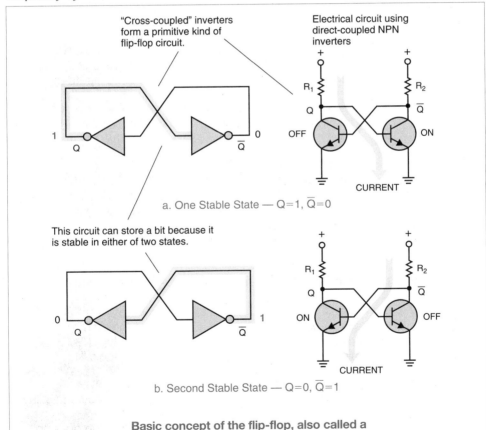

a. One Stable State — Q=1, Q̄=0

b. Second Stable State — Q=0, Q̄=1

Basic concept of the flip-flop, also called a
bistable element or a static memory element.
Source: *Basic Electronics* © 1994, Master Publishing, Inc., Niles, Illinois

E7A05 Which of the following is a circuit that continuously alternates between two states without an external clock?

A. Monostable multivibrator.

B. J-K flip-flop.

C. T flip-flop.

D. Astable multivibrator.

An *astable multivibrator* continuously switches back and forth between *two unstable states without an external clock*. **ANSWER D.**

E7A06 What is a characteristic of a monostable multivibrator?

A. It switches momentarily to the opposite binary state and then returns, after a set time, to its original state.

B. It is a clock that produces a continuous square wave oscillating between 1 and 0.

C. It stores one bit of data in either a 0 or 1 state.

D. It maintains a constant output voltage, regardless of variations in the input voltage.

This type of *multivibrator may momentarily be monostable* (MO MO). It stays in an original state until triggered to its other state, where it remains for a time usually determined by external components, after which it *returns to the original state*. **ANSWER A.**

E7A07 What logical operation does a NAND gate perform?

A. It produces a logic "0" at its output only when all inputs are logic "0."

B. It produces a logic "1" at its output only when all inputs are logic "1."

C. It produces a logic "0" at its output if some but not all of its inputs are logic "1."

D. It produces a logic "0" at its output only when all inputs are logic "1."

The word *NAND* reminds me of the word "naw", meaning no or nothing. The NAND gate *produces a logic "0" at its output only when all inputs are logic "1"*. **ANSWER D.**

E7A08 What logical operation does an OR gate perform?

A. It produces a logic "1" at its output if any or all inputs are logic "1."

B. It produces a logic "0" at its output if all inputs are logic "1."

C. It only produces a logic "0" at its output when all inputs are logic "1."

D. It produces a logic "1" at its output if all inputs are logic "0."

The *OR* gate will produce a *logic "1"* at its output *if any input is, or all inputs are, a logic "1"*. **ANSWER A.**

E7A09 What logical operation is performed by a two-input exclusive NOR gate?

A. It produces a logic "0" at its output only if all inputs are logic "0."

B. It produces a logic "1" at its output only if all inputs are logic "1."

C. It produces a logic "0" at its output if any single input is a logic "1."

D. It produces a logic "1" at its output if any single input is a logic "1."

The *NOR gate* will produce a *logic "0" if any or all inputs are a logic "1"*. With OR gate and NOR gate, spot that word any. **ANSWER C.**

E7A10 What is a truth table?

A. A table of logic symbols that indicate the high logic states of an op-amp.

B. A diagram showing logic states when the digital device's output is true.

C. A list of inputs and corresponding outputs for a digital device.

D. A table of logic symbols that indicates the low logic states of an op-amp.

Logically Speaking of Counters

We use a "*truth table*" in digital circuitry *to characterize a digital device's function*. If you buy the owner's technical manual on a piece of Amateur Radio gear, you will usually find a page describing the truth table of digital devices used in the equipment. **ANSWER C.**

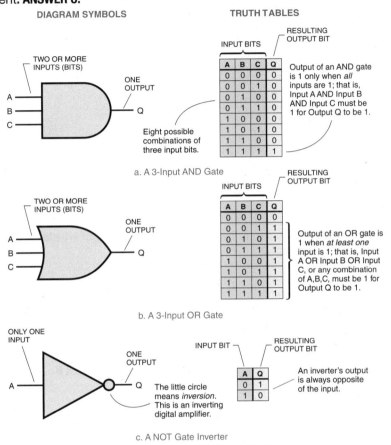

DIAGRAM SYMBOLS

TWO OR MORE INPUTS (BITS)

A
B
C

ONE OUTPUT

Q

Eight possible combinations of three input bits.

TRUTH TABLES

RESULTING OUTPUT BIT

INPUT BITS

A	B	C	Q
0	0	0	0
0	0	1	0
0	1	0	0
0	1	1	0
1	0	0	0
1	0	1	0
1	1	0	0
1	1	1	1

Output of an AND gate is 1 only when *all* inputs are 1; that is, Input A AND Input B AND Input C must be 1 for Output Q to be 1.

a. A 3-Input AND Gate

TWO OR MORE INPUTS (BITS)

A
B
C

ONE OUTPUT

Q

RESULTING OUTPUT BIT

INPUT BITS

A	B	C	Q
0	0	0	0
0	0	1	1
0	1	0	1
0	1	1	1
1	0	0	1
1	0	1	1
1	1	0	1
1	1	1	1

Output of an OR gate is 1 when *at least one* input is 1; that is, Input A OR Input B OR Input C, or any combination of A,B,C, must be 1 for Output Q to be 1.

b. A 3-Input OR Gate

ONLY ONE INPUT

A

ONE OUTPUT

Q

The little circle means *inversion*. This is an inverting digital amplifier.

INPUT BIT

RESULTING OUTPUT BIT

A	Q
0	1
1	0

An inverter's output is always opposite of the input.

c. A NOT Gate Inverter

AND, OR, and NOT Truth Tables

Source: *Basic Electronics* © 1994, Master Publishing, Inc., Niles, Illinois

---

E7A11 What is the name for logic which represents a logic "1" as a high voltage?

A. Reverse Logic.
B. Assertive Logic.
C. Negative logic.
D. Positive Logic.

In a *positive-logic* circuit, the *logic "1" is a high level*, or the most positive level. Think of the four letters in high and the four letters in plus (for positive). **ANSWER D.**

E7A12 What is the name for logic which represents a logic "0" as a high voltage?

A. Reverse Logic.
B. Assertive Logic.
C. Negative logic.
D. Positive Logic.

With the inverting digital amplifier, the NOT gate inverter with an input bit of "1" will result in the inverter's output always opposite, a *"0" as a high voltage*. The letter N is a good way to remember *NOT and negative logic*. **ANSWER C.**

E6C03 Which of the following describes tri-state logic?
A. Logic devices with 0, 1, and high impedance output states.
B. Logic devices that utilize ternary math.
C. Low power logic devices designed to operate at 3 volts.
D. Proprietary logic devices manufactured by Tri-State Devices.

Tri-state logic devices can cut down on the number of individual OR gates needed to route date in a bus wire, where there may be many "listeners." The tri-state device is identified by one additional control lead, and the truth table looks like this for the tri-gate device:

A	Control	B
0	on	0
1	on	1
0	off	high impedance
1	off	high impedance

Tri-state logic devices offer three levels of output: "0," "1," and high impedance.
ANSWER A.

E6C04 Which of the following is the primary advantage of tri-state logic?
A. Low power consumption.
B. Ability to connect many device outputs to a common bus.
C. High speed operation.
D. More efficient arithmetic operations.

Similar to your computer, and even the modern automobile, multiple signals may travel on a pair of wires called a BUS, with multiple sources listening on that BUS for a specific command. To minimize the number of individual gates to accommodate all these signals on the *BUS, tri-state gates are utilized to cut down on the number of components necessary along the BUS.* **ANSWER B.**

E6C12 What is BiCMOS logic?
A. A logic device with two CMOS circuits per package.
B. An FET logic family based on bimetallic semiconductors.
C. A logic family based on bismuth CMOS devices.
D. An integrated circuit logic family using both bipolar and CMOS transistors.

Relatively new to the complementary metal oxide semi conductor is *BiCMOS* logic, *combining MOS* logic *with* increased current capabilities of a *bipolar transistor.* The BiCMOS inverter also offers an increased speed over a stand-alone CMOS gate on a large capacitive load. This increases the current multiplying capability of a bipolar output transistor, with a downside of being more expensive in its intricate BiCMOS construction than individual components that do the same job. **ANSWER D.**

E6C13 Which of the following is an advantage of BiCMOS logic?
- A. Its simplicity results in much less expensive devices than standard CMOS.
- B. It is totally immune to electrostatic damage.
- C. It has the high input impedance of CMOS and the low output impedance of bipolar transistors.
- D. All of these choices are correct.

The *BiCMOS* device offers a *high input impedance* associated with CMOS, yet a *low output impedance* that you would find with a bipolar transistor. It also provides increased gain with low power dissipation in heat. The BiCMOS inverter can be used as a buffer to drive large load capacitances offering increased speed in critical design networks. **ANSWER C.**

E7F01 What is the purpose of a prescaler circuit?
- A. It converts the output of a JK flip-flop to that of an RS flip-flop.
- B. It multiplies a higher frequency signal so a low-frequency counter can display the operating frequency.
- C. It prevents oscillation in a low-frequency counter circuit.
- D. It divides a higher frequency signal so a low-frequency counter can display the input frequency.

You will find a *prescaler* in frequency counters where it is used to *divide down HF and VHF signals* so they can be counted and *displayed on a low-frequency counter*. **ANSWER D.**

E7F02 Which of the following would be used to reduce a signal's frequency by a factor of ten?
- A. A preamp.
- B. A prescaler.
- C. A marker generator.
- D. A flip-flop.

A prescaler has a decade counter that may alter a frequency by a factor of 10. *Prescaler* circuits in frequency counters may *reduce a signal's frequency by a factor of 10* to be displayed within the range of the counter's crystal-controlled circuit. **ANSWER B.**

E7F03 What is the function of a decade counter digital IC?
- A. It produces one output pulse for every ten input pulses.
- B. It decodes a decimal number for display on a seven-segment LED display.
- C. It produces ten output pulses for every input pulse.
- D. It adds two decimal numbers together.

A *decade counter* digital IC gives *one output pulse for every 10 input pulses*. **ANSWER A.**

E7F04 What additional circuitry must be added to a 100-kHz crystal-controlled marker generator so as to provide markers at 50 and 25 kHz?
- A. An emitter-follower.
- B. Two frequency multipliers.
- C. Two flip-flops.
- D. A voltage divider.

We would use *two flip-flops* to take a 100-kHz signal and *provide markers at 50 kHz and 25 kHz*. **ANSWER C.**

E7A13 What is an SR or RS flip-flop?
A. A speed-reduced logic device with high power capability.
B. A set/reset flip-flop whose output is low when R is high and S is low; high when S is high and R is low; and unchanged when both inputs are low.
C. A speed-reduced logic device with very low voltage operation capability.
D. A set/reset flip-flop that toggles whenever the T input is pulsed, unless both inputs are high.

The terms *"SR" or "RS" refer to set/reset*, so the correct answer is narrowed down to B or D. And being a flip-flop, *output will be low when reset is high*, with set being low. *Output will be high when set is high, and reset is low*, and the output will remain unchanged when both inputs are low. **ANSWER B.**

E7A14 What is a JK flip-flop?
A. A flip-flop similar to an RS except that it toggles when both J and K are high.
B. A flip-flop utilizing low power, low temperature Joule-Kelvin devices.
C. A flip-flop similar to a D flip-flop except that it triggers on the negative clock edge.
D. A flip-flop originally developed in Japan and Korea which has very low power consumption.

The JK flip-flop is similar to a reset/set except the *JK toggles when both J and K are high*. **ANSWER A.**

E7A15 What is a D flip-flop?
A. A flip-flop whose output takes on the state of the D input when the clock signal transitions from low to high.
B. A differential class D amplifier used as a flip-flop circuit.
C. A dynamic memory storage element.
D. A flip-flop whose output is capable of both positive and negative voltage excursions.

The D flip-flop will take on the same state on the output as the input, when the clock transitions from LOW to HIGH. **ANSWER A.**

E7F06 What is one purpose of a marker generator?
A. To add audio markers to an oscilloscope.
B. To provide a frequency reference for a phase locked loop.
C. To provide a means of calibrating a receiver's frequency settings.
D. To add time signals to a transmitted signal.

The purpose of a *marker generator* is to provide a means of *calibrating receivers*, as well as showing a calibration mark on monitor scopes and spectrum analyzers. For this question and answer, just look for "calibrating a receiver." **ANSWER C.**

E7F07 What determines the accuracy of a frequency counter?
A. The accuracy of the time base.
B. The speed of the logic devices used.
C. Accuracy of the AC input frequency to the power supply.
D. Proper balancing of the mixer diodes.

When shopping for a *frequency counter*, check out the frequency *time base accuracy*. The stability of the counter will depend on the internal crystal reference, and you should regularly check your crystal reference when calibrating transmit equipment for digital operation with no frequency error greater than 5 Hz. **ANSWER A.**

E7F08 Which of the following is performed by a frequency counter?
- A. Determining the frequency deviation with an FM discriminator.
- B. Mixing the incoming signal with a WWV reference.
- C. Counting the number of input pulses occurring within a specific period of time.
- D. Converting the phase of the measured signal to a voltage which is proportional to the frequency.

There are several manufacturers of excellent amateur radio frequency counters – accurate enough to also be used in the commercial radio service. *A frequency counter counts the number of input pulses in a specific period of time*. You can usually set the sample rate by pushing a button on the front of the counter. **ANSWER C.**

E7F09 What is the purpose of a frequency counter?
- A. To provide a digital representation of the frequency of a signal.
- B. To generate a series of reference signals at known frequency intervals.
- C. To display all frequency components of a transmitted signal.
- D. To provide a signal source at a very accurate frequency.

The frequency counter is a near-field receiver capable of indicating the frequency of a near-field "strongest input" signal that is within the counter's frequency range. Most counters will easily cover from low frequency all the way through 2 GHz ultra-high frequency. The actual purpose of the *frequency counter* is to present *a digital readout of the transmit frequency* you are measuring, or of the frequency of a specific stage within your transceiver when doing maintenance. **ANSWER A.**

Frequency Counter
Courtesy of OptoElectronics

E7F10 What alternate method of determining frequency, other than by directly counting input pulses, is used by some counters?
- A. GPS averaging.
- B. Period measurement plus mathematical computation.
- C. Prescaling.
- D. D/A conversion.

Traditional frequency counters are based on a fixed crystal and multiple prescale stages to count frequencies from DC to GHz. *An alternate method to counting input pulses is to measure the sample frequency in time*, called the period of the cycle, and use this to calculate the frequency. **ANSWER B.**

E7F11 What is an advantage of a period-measuring frequency counter over a direct-count type?

    A. It can run on battery power for remote measurements.

    B. It does not require an expensive high-precision time base.

    C. It provides improved resolution of low-frequency signals within a comparable time period.

    D. It can directly measure the modulation index of an FM transmitter.

A common frequency counter calculates the number of cycles per second (Hertz) of the near-field frequency it is measuring. These low-cost counters are amazingly precise as they average out the readout from internal counters. In the *period-measuring frequency counter*, which is an expensive proposition, the RF wave is divided up into 360 degrees for a *more precise time measurement*. Phase difference comparisons take place in the reference oscillator, leading to improved signal frequency resolution within a comparable time period. This takes place inside many GPS receivers where signals are measured down to 10 to the minus 15th second, the femto second! **ANSWER C.**

Optos and OpAmps Plus Solar

E7G12 What is an integrated circuit operational amplifier?

A. A high-gain, direct-coupled differential amplifier with very high input and very low output impedance.

B. A digital audio amplifier whose characteristics are determined by components external to the amplifier.

C. An amplifier used to increase the average output of frequency modulated amateur signals to the legal limit.

D. An RF amplifier used in the UHF and microwave regions.

The *operational amplifier (op-amp)* is a direct-coupled differential amplifier that offers high gain and high input impedance. The "*differential*" wording refers to the op-amp input design where the output is determined by the difference voltage between the two inputs. Different op-amp circuits and characteristics are determined by the external components connected to the amplifier. This integrated circuit op-amp offers high-gain with a very *high input* impedance, and very *low output impedance*. **ANSWER A.**

E7G01 What primarily determines the gain and frequency characteristics of an op-amp RC active filter?

A. The values of capacitors and resistors built into the op-amp.

B. The values of capacitors and resistors external to the op-amp.

C. The input voltage and frequency of the op-amp's DC power supply.

D. The output voltage and smoothness of the op-amp's DC power supply.

One of the nice features of an op-amp is that it will operate as an RC filter – *resistors (R) and capacitors (C)* are in a circuit outside the op-amp. They are not built into the op-amp – they *are external*. **ANSWER B.**

Operational Amplifier Basics

Ideal Operational Amplifier

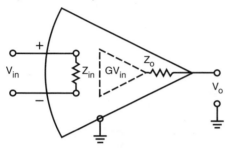

Z_{in} = Infinity

G = Gain = Infinity

Z_o = Zero

Bandwith = Infinity

V_o has no offset
  (V_o=0 when V_{in}=0)

IC Operational Ampliifer

Even though IC operational amplifiers do not meet all the ideal specifications, G and Z_{in} are very large. Because G is very large, any small input voltage would drive the output into saturation (for practical supply voltages); therefore, normal operational amplifier operation is with feedback to set the gain. Here's an example for an inverting amplifier:

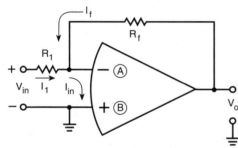

Since Z_{in} is very large, I_{in} = 0

$$\therefore I_1 + I_f = 0 \qquad \therefore \frac{V_{in}}{R_1} + \frac{V_o}{R_f} = 0$$

$$I_1 = \frac{V_{in}}{R_1} \qquad \frac{V_o}{R_f} = -\frac{V_{in}}{R_1}$$

$$I_f = \frac{V_o}{R_f} \qquad \therefore \frac{V_o}{V_{in}} = -\frac{R_f}{R_1}$$

(A) Inverting Input — An input voltage that is more positive on this input will cause the output voltage to be less positive.

(B) Non-Inverting Input — An input voltage that is more positive on this input will cause the output voltage to be more positive.

The gain of an *inverting IC operational amplifier* is:

$$\blacktriangleright\!\!\!\text{E} \quad G = -\frac{R_f}{R_1}$$

The minus sign means the output is out of phase with the input.

E7G02 What is the effect of ringing in a filter?
A. An echo caused by a long time delay.
B. A reduction in high frequency response.
C. Partial cancellation of the signal over a range of frequencies.
D. Undesired oscillations added to the desired signal.

Ringing in a filter may contribute to *unwanted oscillations* that could actually ride along with the desired signal, making it more difficult to copy the signal you want to hear. On Moon bounce, this could be your own echo off the Moon with the added oscillation of a ringing filter. **ANSWER D.**

E7G03 Which of the following is an advantage of using an op-amp instead of LC elements in an audio filter?
A. Op-amps are more rugged.
B. Op-amps are fixed at one frequency.
C. Op-amps are available in more varieties than are LC elements.
D. Op-amps exhibit gain rather than insertion loss.

Here we are, back to that wonderful *op-amp*, which exhibits *high gain with negligible insertion loss* because its input impedance is also high. **ANSWER D.**

E7G04 Which of the following is a type of capacitor best suited for use in high-stability op-amp RC active filter circuits?
A. Electrolytic.
B. Disc ceramic.
C. Polystyrene.
D. Paper.

The tiny *op-amp* filter should use *polystyrene capacitors* because they are very stable and will not vary when the circuit begins to heat up. **ANSWER C.**

E7G05 How can unwanted ringing and audio instability be prevented in a multi-section op-amp RC audio filter circuit?
A. Restrict both gain and Q.
B. Restrict gain, but increase Q.
C. Restrict Q, but increase gain.
D. Increase both gain and Q.

In order to keep an *op-amp* from going into oscillation, *both the gain and the Q are restricted* and set by the feedback circuit. **ANSWER A.**

E7G06 Which of the following is the most appropriate use of an op-amp active filter?
A. As a high-pass filter used to block RFI at the input to receivers.
B. As a low-pass filter used between a transmitter and a transmission line.
C. For smoothing power-supply output.
D. As an audio filter in a receiver.

You will find the *op-amp RC active filter* in the *audio section* of your new transceiver's receiver. As you spend more dollars for that new Extra Class reward radio, filter options become greater, with more filter schemes included in that new, high-priced transceiver. **ANSWER D.**

E7G08 How does the gain of an ideal operational amplifier vary with frequency?
A. It increases linearly with increasing frequency.
B. It decreases linearly with increasing frequency.
C. It decreases logarithmically with increasing frequency.
D. It does not vary with frequency.

The *gain* on an ideal op amp *does not vary with frequency*. **ANSWER D.**

E7G07 What magnitude of voltage gain can be expected from the circuit in Figure E7-4 when R1 is 10 ohms and RF is 470 ohms?

A. 0.21.
B. 94.
C. 47.
D. 24.

Take a look at the operational amplifier basics chart on page 206, and remember the gain formula of an inverting IC op amp:

 $G = -R_F \div R_1 = R_F/R_1.$

The minus sign means the output is out of phase with the input. RF = 470 ohms; R1 = 10 ohms. *Divide 10 into 470, coming out 47.* On a calculator, the keystrokes are: Clear, 470 ÷ 10 = 47. Not really a brain-buster, was it? **ANSWER C.**

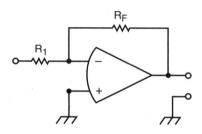

Figure E7-4

E7G09 What will be the output voltage of the circuit shown in Figure E7-4 if R1 is 1000 ohms, RF is 10,000 ohms, and 0.23 volts dc is applied to the input?

A. 0.23 volts.
B. 2.3 volts.
C. -0.23 volts.
D. -2.3 volts.

This time we work the formula: $V_{OUT}/V_{IN} = -R_F/R_1$, when $R_1 = 1000$, and $R_F = 10,000$, the gain of the op amp is $V_{OUT}/V_{IN} = -10$. As a result, the *output = -10 × V_{IN} = -10 × 0.23 = -2.3 volts.* **ANSWER D.**

E7G10 What absolute voltage gain can be expected from the circuit in Figure E7-4 when R1 is 1800 ohms and RF is 68 kilohms?

A. 1.
B. 0.03.
C. 38.
D. 76.

Now we're back to the gain of an inverting IC operational amplifier. *Simply divide 1,800 ohms into 68,000 ohms.* Your calculator keystrokes are: clear, 68000 ÷ 1800 = 37.777, rounded off to your *correct answer of 38.* Simple, huh? **ANSWER C.**

E7G11 What absolute voltage gain can be expected from the circuit in Figure E7-4 when R1 is 3300 ohms and RF is 47 kilohms?

A. 28.
B. 14.
C. 7.
D. 0.07.

One more time, an easy one – *divide* the small number *(3,300) into* the large number *(47,000)*, and you end up with 14.24, with the correct answer of *14.* **ANSWER B.**

E7G13 What is meant by the term op-amp input-offset voltage?
 A. The output voltage of the op-amp minus its input voltage.
 B. The difference between the output voltage of the op-amp and the input voltage required in the immediately following stage.
 C. The differential input voltage needed to bring the open-loop output voltage to zero.
 D. The potential between the amplifier input terminals of the op-amp in an open-loop condition.
If the op amp has been constructed properly, you should see almost *no voltage between the amplifier input terminals* when the feedback loop is closed. **ANSWER C.**

E7G14 What is the typical input impedance of an integrated circuit op-amp?
 A. 100 ohms.
 B. 1000 ohms.
 C. Very low.
 D. Very high.
Remember that the *input impedance* to the ideal *op amp* is always *Hi iN*. **ANSWER D.**

E7G15 What is the typical output impedance of an integrated circuit op-amp?
 A. Very low.
 B. Very high.
 C. 100 ohms.
 D. 1000 ohms.
Remember that the *output impedance* of the ideal op amp is always *low out*.
ANSWER A.

E6F01 What is photoconductivity?
 A. The conversion of photon energy to electromotive energy.
 B. The increased conductivity of an illuminated semiconductor.
 C. The conversion of electromotive energy to photon energy.
 D. The decreased conductivity of an illuminated semiconductor .
The *photocell* has high resistance when no light shines on it, and a *varying resistance when light shines on it*. It can be used in an ON or OFF state as a simple beam alarm across a doorway, or may be found in almost all modern SLR cameras as a sensitive light meter to judge the amount of light present. **ANSWER B.**

E6F02 What happens to the conductivity of a photoconductive material when light shines on it?
 A. It increases.
 B. It decreases.
 C. It stays the same.
 D. It becomes unstable.
In the dark, a photocell has high resistance. When you *shine light on it*, the *conductivity increases* as the *resistance decreases*. **ANSWER A.**

E6F03 What is the most common configuration of an optoisolator or optocoupler?
 A. A lens and a photomultiplier.
 B. A frequency modulated helium-neon laser.
 C. An amplitude modulated helium-neon laser.
 D. An LED and a phototransistor.
Optocouplers are found in many modern, high-frequency transceivers. When you spin the dial to tune, you are not turning a giant capacitor, but rather interrupting an *LED* light beam on a *phototransistor*. **ANSWER D.**

E6F08 Why are optoisolators often used in conjunction with solid state circuits when switching 120 VAC?

A. Optoisolators provide a low impedance link between a control circuit and a power circuit.

B. Optoisolators provide impedance matching between the control circuit and power circuit.

C. Optoisolators provide a very high degree of electrical isolation between a control circuit and the circuit being switched.

D. Optoisolators eliminate the effects of reflected light in the control circuit.

The chief advantage of the *optoisolator* at household voltages is to provide a *very high degree of electrical isolation between* the control circuit and the house power circuit. Look for the answer "isolation" to agree with the optoisolator device. **ANSWER C.**

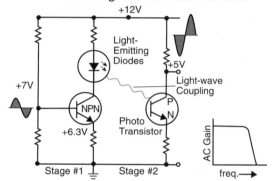

Optical Coupling
Source: *Basic Communications Electronics*, Hudson & Luecke, © 1999
Master Publishing, Inc., Niles, IL

E6F05 Which of the following describes an optical shaft encoder?

A. A device which detects rotation of a control by interrupting a light source with a patterned wheel.

B. A device which measures the strength a beam of light using analog to digital conversion.

C. A digital encryption device often used to encrypt spacecraft control signals.

D. A device for generating RTTY signals by means of a rotating light source.

When you turn the big VFO knob of that new high frequency transceiver that you awarded yourself for passing the Extra Class, an *LED light source shining through* clear and black *window patterns on a wheel* will detect rotation of a control, with a logic circuit counting the number of light pulses making it through to the *optical shaft encoder*. **ANSWER A.**

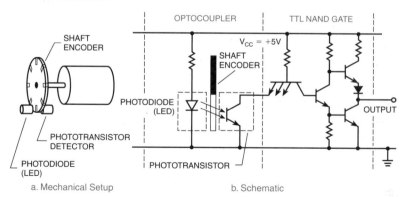

Optocoupler Used for Shaft Encoder

E6F04 What is the photovoltaic effect?
A. The conversion of voltage to current when exposed to light.
B. The conversion of light to electrical energy.
C. The conversion of electrical energy to mechanical energy.
D. The tendency of a battery to discharge when used outside.

The term *"photovoltaic"* refers to *light photons being converted to electrical energy*.
ANSWER B.

E6F09 What is the efficiency of a photovoltaic cell?
A. The output RF power divided by the input dc power.
B. The effective payback period.
C. The open-circuit voltage divided by the short-circuit current under full illumination.
D. The relative fraction of light that is converted to current.

Lead sulfide offers good efficiency in a photovoltaic cell – lead takes in the infrared and cadmium takes in visible light, with *solar cell efficiency* as t*he relative fraction of light that ultimately gets converted to current*. **ANSWER D.**

E6F06 Which of these materials is affected the most by photoconductivity?
A. A crystalline semiconductor.   C. A heavy metal.
B. An ordinary metal.   D. A liquid semiconductor.

The *crystalline semiconductor* is found inside the optical shaft encoder mechanism, and might also be found outside your ham shack as part of your solar-charging system. The light-emitting diode is normally the light source inside your tuning dial optocoupler. **ANSWER A.**

Resistance Variation of a Photodiode			
Light Incidence	**Current I**	**Resistance**	**Test Conditions**
Dark	5 nA	2000 M	$V_R = 10$ V *$E_e = 0$
Light	15 µA	666.7 k	$V_R = 10$ V $E_e = 250$ µW/cm^2 at 940 nm
*Irradiance (E_e) is the radiant power per unit area incident on surface			

E6F11 Which of the following is the approximate open-circuit voltage produced by a fully-illuminated silicon photovoltaic cell?
A. 0.1 V.   C. 1.5V.
B. 0.5 V.   D. 12 V.

That new *solar panel* you purchased for your Field Day ham radio battery likely has an output of 17 volts with no load. Add it to a fresh battery and it drops down to about 12 or 13 volts under load. But in this question they ask about a single cell, and *each cell is rated at 0.5 volts*. CELL, not panel voltage. **ANSWER B.**

E6F12 What absorbs the energy from light falling on a photovoltaic cell?

A. Protons.

B. Photons.

C. Electrons.

D. Holes.

Photon energy from sunlight is absorbed by the electrons in the photovoltaic cell, creating voltage output from sunlight. When sunlight photons strike the panel, an *electron will absorb the energy* and develop a new energy level, *creating the photovoltaic process*. **ANSWER C.**

E6F10 What is the most common type of photovoltaic cell used for electrical power generation?

A. Selenium.

B. Silicon.

C. Cadmium Sulfide.

D. Copper oxide.

The *solar cell* is made up of *silicon wafers*, which offer relatively high voltage output and durability for the many years it will be out in the sunlight. **ANSWER B.**

Solar Panel array for Charging Storage Batteries

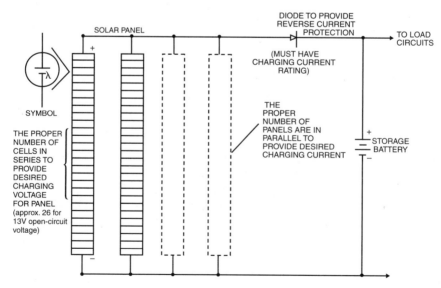

Schematic of Solar Panel for Charging Storage Battery

Test Gear, Testing, Testing 1, 2, 3

MY SCOPE SAYS MY SIGNAL MATCHES MY FLAT-TOP!!

E4A01 How does a spectrum analyzer differ from an oscilloscope?

A. A spectrum analyzer measures ionospheric reflection; an oscilloscope displays electrical signals.

B. A spectrum analyzer displays the peak amplitude of signals; an oscilloscope displays the average amplitude of signals.

C. A spectrum analyzer displays signals in the frequency domain; an oscilloscope displays signals in the time domain.

D. A spectrum analyzer displays radio frequencies; an oscilloscope displays audio frequencies.

The *spectrum analyzer displays* the strength of signals at particular frequencies according to frequency (*in the frequency domain*) along a horizontal axis. A signal can be locked in and centered in the middle of the screen with 25 kHz to the left and to the right for a close examination of any off-frequency spurs it might have. A simple oscilloscope cannot display the signal in the frequency domain – it displays the signal on a time axis (in the time domain) going from left to right. With your output signal displayed on a spectrum analyzer, you may adjust your transmitter output network for minimum spurious signals. **ANSWER C.**

Spectrum Analyzer
Courtesy of Hewlett-Packard Company

E4A02 Which of the following parameters would a spectrum analyzer display on the horizontal axis?

A. SWR.

B. Q.

C. Time.

D. Frequency.

The *spectrum analyzer* displays a portion of the *frequency* spectrum and shows you the amplitude of signals within that spectrum. The *horizontal axis* represents the frequency of the signals, while the vertical axis represents their amplitudes. **ANSWER D.**

E4A03 Which of the following parameters would a spectrum analyzer display on the vertical axis?

A. Amplitude.

B. Duration.

C. SWR.

D. Q.

The *vertical axis* of the spectrum analyzer displays the *amplitude* of the signal. **ANSWER A.**

E4A04 Which of the following test instruments is used to display spurious signals from a radio transmitter?

A. A spectrum analyzer.

B. A wattmeter.

C. A logic analyzer.

D. A time-domain reflectometer.

You can check for *spurious signals* using a *spectrum analyzer.* **ANSWER A.**

E4A05 Which of the following test instruments is used to display intermodulation distortion products in an SSB transmission?

A. A wattmeter.

B. A spectrum analyzer.

C. A logic analyzer.

D. A time-domain reflectometer.

Only the *spectrum analyzer* allows you to see your output versus frequency by displaying it in the frequency domain. This allows you to check for *harmonics* that cause *distortion* in your SSB transmitter. **ANSWER B.**

E4A06 Which of the following could be determined with a spectrum analyzer?

A. The degree of isolation between the input and output ports of a 2 meter duplexer.

B. Whether a crystal is operating on its fundamental or overtone frequency.

C. The spectral output of a transmitter.

D. All of these choices are correct .

As an Extra Class operator, the spectrum analyzer may become your ultimate best radio service monitor. The *spectrum analyzer* allows you to view the purity of your transmitted signal. The spectrum analyzer also can show fundamental and overtone crystal response, and you could use it for tuning-up the input and output ports of a VHF or UHF duplexer. *It can do all of these things!* With the addition of an attenuator to protect the spectrum analyzer from overload, isolation of the transmit frequency in the receive port of a duplexer can be determined. **ANSWER D.**

E4A12 Which of the following procedures is an important precaution to follow when connecting a spectrum analyzer to a transmitter output?

A. Use high quality double shielded coaxial cables to reduce signal losses.

B. Attenuate the transmitter output going to the spectrum analyzer.

C. Match the antenna to the load.

D. All of these choices are correct.

Never transmit directly into a *spectrum analyzer*. You'll blow it up, for sure. Your *transmitter signal must be attenuated* to just a fraction of a watt in order to protect the spectrum analyzer input. I make it a practice to NEVER direct couple to a spectrum analyzer. Rather, I use a little pigtail and "sniff" my transmit RF. **ANSWER B.**

E4B02 What is an advantage of using a bridge circuit to measure impedance?
   A. It provides an excellent match under all conditions.
   B. It is relatively immune to drift in the signal generator source.
   C. The measurement is based on obtaining a signal null, which can be done very precisely.
   D. It can display results directly in Smith chart format.

Multiband, high-frequency antenna traps, containing both coil inductive reactance and outer sleeve capacitive reactance, can be tuned to a specific frequency that will stop the RF signal from going further out on the element. A portable impedance-measuring bridge circuit will show a much more precise dip, or null, in voltage, as opposed to many other methods of impedance measurement. Try to loosely couple the impedance *bridge circuit* so the *pronounced dip* is as *precisely on frequency* as possible. **ANSWER C.**

E4B14 What happens if a dip meter is too tightly coupled to a tuned circuit being checked?
   A. Harmonics are generated.   C. Cross modulation occurs.
   B. A less accurate reading results.   D. Intermodulation distortion occurs.

If you put your *dip meter right on top of a coil*, you will have a *less accurate reading* – only get close enough to see a smooth dip on your readout. On the SWR analyzer type of dip meters, everything is done for you automatically inside the box. All you need to do is to screw on the antenna cable and sweep the frequency dial to see where the antenna is resonant. **ANSWER B.**

E4B01 Which of the following factors most affects the accuracy of a frequency counter?
   A. Input attenuator accuracy.   C. Decade divider accuracy.
   B. Time base accuracy.   D. Temperature coefficient of the logic.

Everyone should have *frequency counter* in their ham shack. Be sure to get one that has *excellent time base*. **ANSWER B.**

E4B05 If a frequency counter with a specified accuracy of 10 ppm reads 146,520,000 Hz, what is the most the actual frequency being measured could differ from the reading?
   A. 146.52 Hz.   C. 146.52 kHz.
   B. 10 Hz.   D. 1465.20 Hz.

Keystroke the following on your calculator: Clear *146.52* × *10 = 1465.2 Hz. You are multiplying the frequency in MHz by the time base accuracy of 10 ppm*. **ANSWER D.**

Frequency Counter
Courtesy of OptoElectronics

E4B03 If a frequency counter with a specified accuracy of +/- 1.0 ppm reads 146,520,000 Hz, what is the most the actual frequency being measured could differ from the reading?

A. 165.2 Hz.
B. 14.652 kHz.
C. 146.52 Hz.
D. 1.4652 MHz.

You may have a question on your test regarding the error readout on a frequency counter with a specific part per million (ppm) time base. It's easy to solve these problems with a simple calculator. Since they're asking parts per million, first convert the frequency given in Hz to MHz. You do this simply by moving the decimal point 6 places to the left. Now multiply the frequency in MHz times the parts per million time base accuracy (in this case 1 ppm). This will give you the error they are looking for in the answer. Try this on your calculator:
Clear *146.52 × 1 = 146.52 Hz.* **ANSWER C.**

Counter Readout Error
Readout Error = f × a

f = Frequency in **MHz** being measured
a = Counter accuracy in **parts per million**

E4B04 If a frequency counter with a specified accuracy of +/- 0.1 ppm reads 146,520,000 Hz, what is the most the actual frequency being measured could differ from the reading?

A. 14.652 Hz.
B. 0.1 MHz.
C. 1.4652 Hz.
D. 1.4652 kHz.

Hertz to MHz will give you a frequency of 146.52. Multiple that by 0.1 parts per million, and see if your calculator reads 14.652 Hz. The keystrokes are: Clear *146.52 × 0.1 = 14.652.* **ANSWER A.**

E4B06 How much power is being absorbed by the load when a directional power meter connected between a transmitter and a terminating load reads 100 watts forward power and 25 watts reflected power?

A. 100 watts.
B. 125 watts.
C. 25 watts.
D. 75 watts.

Here's a great question that points out the importance of working your antenna system for maximum power transfer, minimum SWR, and a better signal on the air than simply using an antenna tuner to make up for manually working over the antenna elements for better antenna resonance. *If 25 watts are reflected from 100 watts initially delivered to the load (the antenna), only 75 watts are truly being absorbed by the antenna.* The tuner won't help; it is just fooling your transceiver into thinking everything is okay with your antenna system! The reflected 25 watts does not represent lost power, except for line loss attenuation over and above the matched attenuation of the feed line. What is being absorbed by the load and radiated will be 75 watts, with those extra 25 watts circulating in the feed line between the transmitter and the load. Good reading is Maxell's Reflections II. **ANSWER D.**

E4B07 Which of the following is good practice when using an oscilloscope probe?

A. Keep the signal ground connection of the probe as short as possible.
B. Never use a high impedance probe to measure a low impedance circuit.
C. Never use a DC-coupled probe to measure an AC circuit.
D. All of these choices are correct.

When working on radio equipment using an *oscilloscope probe*, we want the *probe's* signal ground lead to be *kept as short as possible* and clipped to the ground chassis near where you are testing. The short ground lead will minimize stray pickup from other stages of the radio. **ANSWER A.**

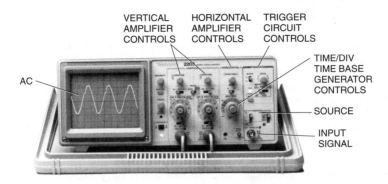

VERTICAL AMPLIFIER CONTROLS

HORIZONTAL AMPLIFIER CONTROLS

TRIGGER CIRCUIT CONTROLS

TIME/DIV TIME BASE GENERATOR CONTROLS

SOURCE

INPUT SIGNAL

AC

A Typical Oscilloscope showing an AC Waveform
Source: *Basic Electronics* © 1994, Master Publishing, Inc., Niles, Illinois

E4B13 How is the compensation of an oscilloscope probe typically adjusted?

A. A square wave is displayed and the probe is adjusted until the horizontal portions of the displayed wave are as nearly flat as possible.

B. A high frequency sine wave is displayed and the probe is adjusted for maximum amplitude.

C. A frequency standard is displayed and the probe is adjusted until the deflection time is accurate.

D. A DC voltage standard is displayed and the probe is adjusted until the displayed voltage is accurate.

If you regularly use an oscilloscope when working on your ham radio equipment, and you just purchased a new probe, use your *square wave* generator and adjust the oscilloscope for the horizontal portion of the displayed wave *to be as flat as possible*. **ANSWER A.**

E4B08 Which of the following is a characteristic of a good DC voltmeter?

A. High reluctance input.

B. Low reluctance input.

C. High impedance input.

D. Low impedance input.

Your modern high-frequency or VHF/UHF transceiver has circuits so sensitive that they will quit oscillating if you attempt to measure *DC voltage* with an inexpensive, low-impedance multimeter. *High impedance input* – usually greater than 25 k ohms per volt – may help to prevent loading or changing the characteristics of the circuit. An excellent example is your digital multimeter with an input impedance of 10 meg-ohms. **ANSWER C.**

Digital Multitester
Source: *Using Your Meter* © 2012,
Master Publishing, Inc., Niles, Illinois

E4B09 What is indicated if the current reading on an RF ammeter placed in series with the antenna feed line of a transmitter increases as the transmitter is tuned to resonance?

    A. There is possibly a short to ground in the feed line.
    B. The transmitter is not properly neutralized.
    C. There is an impedance mismatch between the antenna and feed line.
    D. There is more power going into the antenna

When I am working aboard boats, testing the output of a long-wire antenna tuner, I insert a series thermocouple RF ammeter to monitor RF output levels. This lets me confirm that the tuner is operating properly when I change bands, and see *RF power increase* up to an amp *going into the long-wire antenna*. Another good indicator of RF output to the antenna circuit is the common fluorescent tube. If the tube glows on CW dots when held within an inch of the antenna radiating element on high frequency, it is a good indication that there is plenty of RF to work the world. **ANSWER D.**

E4B12 What is the significance of voltmeter sensitivity expressed in ohms per volt?

    A. The full scale reading of the voltmeter multiplied by its ohms per volt rating will provide the input impedance of the voltmeter.
    B. When used as a galvanometer, the reading in volts multiplied by the ohms/volt will determine the power drawn by the device under test.
    C. When used as an ohmmeter, the reading in ohms divided by the ohms/volt will determine the voltage applied to the circuit.
    D. When used as an ammeter, the full scale reading in amps divided by ohms/volt will determine the size of shunt needed.

When measuring anything in a circuit, it is important that the measuring instrument cause as little change to the circuit as possible. Since a voltmeter is placed across a circuit (connected between two points in that circuit), it acts as a resistance in parallel with that part of the circuit. In order to cause the least change, the meter should be as high a resistance as possible. *By multiplying the ohms per volt rating times the maximum voltage on a given scale of the meter, you can determine the equivalent resistance your meter will look like to the circuit.* **ANSWER A.**

E4B10 Which of the following describes a method to measure intermodulation distortion in an SSB transmitter?

    A. Modulate the transmitter with two non-harmonically related radio frequencies and observe the RF output with a spectrum analyzer.
    B. Modulate the transmitter with two non-harmonically related audio frequencies and observe the RF output with a spectrum analyzer.
    C. Modulate the transmitter with two harmonically related audio frequencies and observe the RF output with a peak reading wattmeter.
    D. Modulate the transmitter with two harmonically related audio frequencies and observe the RF output with a logic analyzer.

When looking at your signal on a spectrum analyzer, a whistle into the microphone is not a good test of transmit linearity. *Two non-harmonically related audio tones* from a 2 tone generator is a much better way to observe the RF output on your new $10,000 spectrum analyzer. **ANSWER B.**

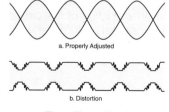
a. Properly Adjusted

b. Distortion

Two-Tone Test

Antennas

GIT ALONG,
LITTLE YAGI!

E9A15 What is meant by the radiation resistance of an antenna?
    A. The combined losses of the antenna elements and feed line.
    B. The specific impedance of the antenna.
    C. The value of a resistance that would dissipate the same amount of power as that radiated from an antenna.
    D. The resistance in the atmosphere that an antenna must overcome to be able to radiate a signal.

Radiation resistance is an important term because it compares the power radiated from an antenna to the equivalent resistance that would dissipate the same amount of power as heat. The higher the radiation resistance of an antenna, the better its performance. Don't confuse radiation resistance with simple DC resistance – they are exactly opposite. Radiation resistance is *the equivalent resistance that would dissipate the same amount of power as that radiated from the antenna.* **ANSWER C.**

E9A06 What is included in the total resistance of an antenna system?
    A. Radiation resistance plus space impedance.
    B. Radiation resistance plus transmission resistance.
    C. Transmission-line resistance plus radiation resistance.
    D. Radiation resistance plus ohmic resistance.

To calculate the *total resistance* of an antenna system, *add radiation resistance plus all ohmic resistances* within the coils and conductors. **ANSWER D.**

E9A05 Which of the following factors may affect the feed point impedance of an antenna?
A. Transmission line length.
B. Antenna height, conductor length/diameter ratio, and location of nearby conductive objects.
C. Constant feed point impedance.
D. Sunspot activity and time of day.

You just built that new beam, and the three largest factors that will affect the *feed point impedance* of that antenna are getting it more than a half wavelength in *height*, the *element length to diameter ratio*, and *how close it is to a nearby conductive object* (like another band beam below it). **ANSWER B.**

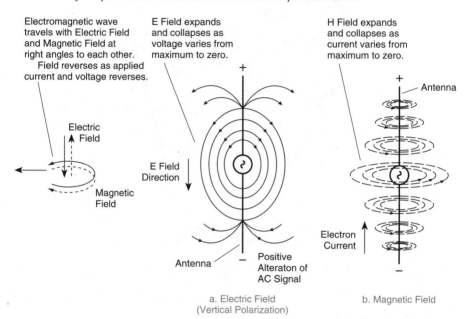

Electromagnetic wave travels with Electric Field and Magnetic Field at right angles to each other. Field reverses as applied current and voltage reverses.

E Field expands and collapses as voltage varies from maximum to zero.

H Field expands and collapses as current varies from maximum to zero.

Electric Field

Magnetic Field

E Field Direction

Antenna

Positive Alteraton of AC Signal

Antenna

Electron Current

a. Electric Field (Vertical Polarization)

b. Magnetic Field

Propagating Electromagnetic Waves

Source: *Antennas – Selection and Installation*, A.J. Evans, Copyright ©1986 Master Publishing, Inc., Niles, Illinois

---

E9A04 Why would one need to know the feed point impedance of an antenna?
A. To match impedances in order to minimize standing wave ratio on the transmission line.
B. To measure the near-field radiation density from a transmitting antenna.
C. To calculate the front-to-side ratio of the antenna.
D. To calculate the front-to-back ratio of the antenna.

When the antenna (the load) *has the same characteristic impedance as the feed line, maximum power will be transferred*. A mismatch may result in high SWR, transmit power feedback, and potential transmitter heat-up. Most coaxial cable presents a 50-ohm impedance. **ANSWER A.**

E9A10 How is antenna efficiency calculated?
    A. (radiation resistance / transmission resistance) × 100%.
    B. (radiation resistance / total resistance) × 100%.
    C. (total resistance / radiation resistance) × 100%.
    D. (effective radiated power / transmitter output) × 100%.

The efficiency of your *antenna* is an important consideration when you plan to upgrade your antenna system. *Efficiency equals the radiation resistance divided by the total resistance, multiplied by 100 percent*. You want to keep your ohmic resistance as low as possible to make your antenna more efficient. Generally, bigger antenna elements have greater radiation resistance, lower ohmic resistance, higher efficiency, and increased bandwidth. Bigger is always better! **ANSWER B.**

E9A11 Which of the following choices is a way to improve the efficiency of a ground-mounted quarter-wave vertical antenna?
    A. Install a good radial system.
    B. Isolate the coax shield from ground.
    C. Shorten the radiating element.
    D. Reduce the diameter of the radiating element.

Most *vertical antennas* are one-quarter wavelength long, and use a mirror image counterpoise to compare to that of a halfwave antenna. A good *ground radial system* will give your signal more punch, lower the noise floor, lower your take-off angle, and give you good DX. **ANSWER A.**

E9A12 Which of the following factors determines ground losses for a ground-mounted vertical antenna operating in the 3-30 MHz range?
    A. The standing-wave ratio.
    B. Distance from the transmitter.
    C. Soil conductivity.
    D. Take-off angle.

Dirt is ground, right? Likely not – dry *soil* is not nearly as effective a *conductor* as 4 quarter-wavelength radials per band on that new HF vertical antenna in your back yard. Unless you live in a marsh, you'll probably increase antenna performance by mounting it up on the roof, and creating your own ground plane radial system. Looking to really increase your 5 band trap vertical performance on HF? Replace those quarter wavelength ground wires with 3 inch wide copper foil strips, 4 per band, each a quarter wavelength long. **ANSWER C.**

E9C11 How is the far-field elevation pattern of a vertically polarized antenna affected by being mounted over seawater versus rocky ground?
    A. The low-angle radiation decreases.
    B. The high-angle radiation increases.
    C. Both the high- and low-angle radiation decrease.
    D. The low-angle radiation increases.

Here we go – mount it over *salt water* and achieve a *low-angle radiation increase*. This gives you longer-range skywave signals. **ANSWER D.**

E9C13 What is the main effect of placing a vertical antenna over an imperfect ground?

    A. It causes increased SWR.
    B. It changes the impedance angle of the matching network.
    C. It reduces low-angle radiation.
    D. It reduces losses in the radiating portion of the antenna.

When I outfit boaters with their high frequency antenna systems, I always couple the antenna ground system to sea water. Nothing beats a direct contact to conductive sea water for the antenna ground, lowering the takeoff angle for maximum DX. *Mounting a vertical antenna on your roof*, relying only on the metal vent pipe as the ground plane, *will produce multiple near-vertical radiation angles, and can reduce the DX low-angle radiation* capability you were hoping to achieve. Any vertical antenna needs a major ground system below it if you plan to work more than the lower 48! **ANSWER C.**

E9D14 Which of the following types of conductor would be best for minimizing losses in a station's RF ground system?

    A. A resistive wire, such as a spark plug wire.
    B. A wide flat copper strap.
    C. A cable with 6 or 7 18-gauge conductors in parallel.
    D. A single 12 or 10-gauge stainless steel wire.

What happens when you pass RF through a coil? The coil has inductive reactance, and at certain frequencies the impedance will be high. For ground systems to be effective, we want to minimize any wire runs that will look like a coil and create unwanted impedance. *Copper ground foil is our choice for good grounding* techniques, and it is available in 25 foot, 50 foot, and custom-length rolls from your local ham radio dealer or a copper foil manufacturer. Three inch wide foil is my favorite for minimum inductive reactance. **ANSWER B.**

E9D15 Which of the following would provide the best RF ground for your station?

    A. A 50-ohm resistor connected to ground.
    B. An electrically-short connection to a metal water pipe.
    C. An electrically-short connection to 3 or 4 interconnected ground rods driven into the Earth.
    D. An electrically-short connection to 3 or 4 interconnected ground rods via a series RF choke.

A good earth ground to your new $10,000 Extra Class-reward transceiver should be *copper foil connected to three or four ground rods, driven in to the earth*. Follow manufacturer suggestions on how to ground your tower, too. **ANSWER C.**

E9C01 What is the radiation pattern of two 1/4-wavelength vertical antennas spaced 1/2-wavelength apart and fed 180 degrees out of phase?

    A. A cardioid.
    B. Omnidirectional.
    C. A figure-8 broadside to the axis of the array.
    D. A figure-8 oriented along the axis of the array.

Two quarter-wave vertical antennas spaced *one-half wavelength* apart and *fed 180 degrees out of phase* results in a *figure-8 radiation pattern* with best radiation in line with the antennas. **ANSWER D.**

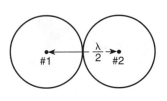

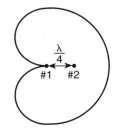

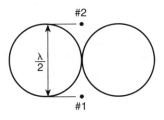

| a. Two λ/4 Verticals Fed 180° Out of Phase (Figure 8 End-Fire In-Line) | b. Two λ/4 Verticals Fed 90° Out of Phase (Cardioid) | c. Two λ/4 Verticals Fed In Phase (Figure 8 Broadside) |

Antenna Radiation Patterns

E9C02 What is the radiation pattern of two 1/4-wavelength vertical antennas spaced 1/4-wavelength apart and fed 90 degrees out of phase?

A. A cardioid.
B. A figure-8 end-fire along the axis of the array.
C. A figure-8 broadside to the axis of the array.
D. Omnidirectional.

Remember when we talked about RDF antennas for direction finding? *Two quarter-wave verticals* spaced *one-quarter wavelength apart* and *fed 90 degrees out of phase* results in a *cardioid pattern*. This is called "unidirectional." **ANSWER A.**

E9C03 What is the radiation pattern of two 1/4-wavelength vertical antennas spaced 1/2-wavelength apart and fed in phase?

A. Omnidirectional.
B. A cardioid.
C. A Figure-8 broadside to the axis of the array.
D. A Figure-8 end-fire along the axis of the array.

Two quarter-waves spaced *one-half wavelength apart* and *fed in phase* will also result in a *figure-8 pattern*, but maximum gain will be *broadside to the antennas*. **ANSWER C.**

E9A07 What is a folded dipole antenna?

A. A dipole one-quarter wavelength long.
B. A type of ground-plane antenna.
C. A dipole constructed from one wavelength of wire forming a very thin loop.
D. A dipole configured to provide forward gain.

The folded dipole is a fun one to build out of television twin-lead cable. As shown in the diagram, the twin-lead *folded dipole is constructed by using a half-wavelength piece of twin lead to form a loop that is one full wavelength long*. Scrape the insulation from the wire ends and twist the wires together on each end. In the center of one of the leads in the twin-lead, cut the lead and scrape some of the insulation off to expose bare wires on the end of each quarter-wave piece. Feed these wires with standard 300-ohm twin-lead, or connect a balun transformer to the bare wires to permit 50-ohm or 75-ohm coax cable to feed the folded dipole. **ANSWER C.**

E9D10 What is the approximate feed point impedance at the center of a two-wire folded dipole antenna?

A. 300 ohms.
B. 72 ohms.
C. 50 ohms.
D. 450 ohms.

The *impedance at the center of a folded dipole* may be a whopping *300 ohms*. You'll need an impedance-matching transformer to feed it with 50-ohm coax; however, ordinary TV twin-lead would match very well. **ANSWER A.**

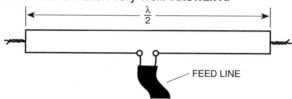

Folded Dipole Antenna Made from 300-Ohm Twin-Lead

E9A01 Which of the following describes an isotropic antenna?

A. A grounded antenna used to measure earth conductivity.
B. A horizontally polarized antenna used to compare Yagi antennas.
C. A theoretical antenna used as a reference for antenna gain.
D. A spacecraft antenna used to direct signals toward the Earth.

An *isotropic radiator* is simply a theoretical antenna; and in every April 1st issue of many magazines, advertisers try to sell "isotropic radiators" to the unsuspecting ham! We use it as a *theoretical reference to calculate the gain* or loss of real antenna systems. **ANSWER C.**

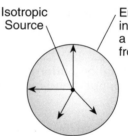

Isotropic Source — Energy radiates equally in all directions forming a sphere of radiation from the point source.

Isotropic Radiator Pattern

E9A03 Which of the following antennas has no gain in any direction?

A. Quarter-wave vertical.
B. Yagi.
C. Half-wave dipole.
D. Isotropic antenna.

It is the theoretical *isotropic radiator*, on paper, that has equal signal dispersion in all directions, and *no gain in any direction*. **ANSWER D.**

E9A02 How much gain does a 1/2-wavelength dipole in free space have compared to an isotropic antenna?

A. 1.55 dB.
B. 2.15 dB.
C. 3.05 dB.
D. 4.30 dB.

The *halfwave dipole* will have approximately *2.15 dB gain* over the theoretical isotropic radiator. **ANSWER B.**

E9A13 How much gain does an antenna have compared to a 1/2-wavelength dipole when it has 6 dB gain over an isotropic antenna?

A. 3.85 dB.
B. 6.0 dB.
C. 8.15 dB.
D. 2.79 dB.

The halfwave dipole has about 2.15 dB gain over an isotropic radiator. If an antenna under measurement has 6 dB gain over an isotropic radiator, *6 dB − 2.15 dB = about 3.85 dB gain over the dipole.* **ANSWER A.**

E9A14 How much gain does an antenna have compared to a 1/2-wavelength dipole when it has 12 dB gain over an isotropic antenna?

A. 6.17 dB.   C. 12.5 dB.

B. 9.85 dB.   D. 14.15 dB.

Here is a bigger antenna that has 12 dB gain over an isotropic radiator, so *12 dB gain − 2.15 dB = about 9.85 dB over the dipole*. **ANSWER B.**

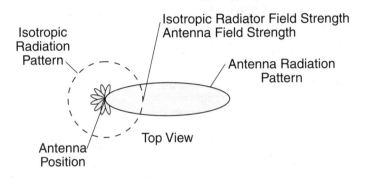

"Gain" of an Antenna

E9D11 What is the function of a loading coil as used with an HF mobile antenna?

A. To increase the SWR bandwidth.   C. To lower the Q.

B. To lower the losses.   D. To cancel capacitive reactance.

Since X_L must equal X_C in order for a whip antenna to achieve resonance, we need to *add a loading coil* on an antenna for lower frequency operation to *tune out capacitive reactance*. **ANSWER D.**

E9D08 What happens to the bandwidth of an antenna as it is shortened through the use of loading coils?

A. It is increased.

B. It is decreased.

C. No change occurs.

D. It becomes flat.

When you *shorten a vertical antenna with a loading coil, bandwidth decreases*, and losses increase. **ANSWER B.**

E9A09 What is meant by antenna bandwidth?

A. Antenna length divided by the number of elements.

B. The frequency range over which an antenna satisfies a performance requirement.

C. The angle between the half-power radiation points.

D. The angle formed between two imaginary lines drawn through the element ends.

If you operate on the lower bands, such as 40 meters, antenna bandwidth is an important consideration. *Bandwidth is the frequency range over which you can expect your antenna to perform well*. Small antennas on 40 meters don't have a very wide bandwidth. Monster antennas for 40 meters have good bandwidth. **ANSWER B.**

E9D06 Why should an HF mobile antenna loading coil have a high ratio of reactance to resistance?

    A. To swamp out harmonics.
    B. To maximize losses.
    C. To minimize losses.
    D. To minimize the Q.

When you see those giant open-air coils halfway up the mast on a mobile whip antenna, you are looking at *high Q coils which minimize losses*. The bigger the coil, the lower the losses. **ANSWER C.**

E9D05 Where should a high-Q loading coil be placed to minimize losses in a shortened vertical antenna?

    A. Near the center of the vertical radiator.
    B. As low as possible on the vertical radiator.
    C. As close to the transmitter as possible.
    D. At a voltage node.

To minimize coil losses and to increase performance of a loaded vertical antenna, the *loading coil* should be near the *center of the vertical radiator*, not at the base. **ANSWER A.** ☞ **Visit: www.hy-gain.com**

Mobile Antenna Loading Coils

E9D09 What is an advantage of using top loading in a shortened HF vertical antenna?

    A. Lower Q.   C. Higher losses.
    B. Greater structural strength.   D. Improved radiation efficiency.

We use *top loading* and helical loading on a shortened HF vertical antenna to *improve radiation efficiency*. With this in mind, it's easy to understand why a little 96-inch CB whip antenna with an automatic antenna tuner inside the trunk of your vehicle will not perform as well as a resonant center-loaded, helical-loaded, or top-loaded vertical antenna. **ANSWER D.**

E9D13 What happens to feed point impedance at the base of a fixed-length HF mobile antenna as the frequency of operation is lowered?

    A. The radiation resistance decreases and the capacitive reactance decreases.
    B. The radiation resistance decreases and the capacitive reactance increases.
    C. The radiation resistance increases and the capacitive reactance decreases.
    D. The radiation resistance increases and the capacitive reactance increases.

As we *operate lower on the bands*, the base feed *point resistance decreases* (lower decreases), and the *capacitive reactance (X_C) increases*. You can visualize the correct answer by thinking the lower you operate in frequency, the more coil (inductive reactance, X_L) you will need to offset and equal the added capacitive reactance, which increases. **ANSWER B.**

E9D12 What is one advantage of using a trapped antenna?
A. It has high directivity in the higher-frequency bands.
B. It has high gain.
C. It minimizes harmonic radiation.
D. It may be used for multiband operation.

The *trap antenna* may be used for *multiband operation*, from a single feed line. There are trap dipoles, trap verticals, and trap beam antennas for multiband operation. **ANSWER D.**

E9D07 What is a disadvantage of using a multiband trapped antenna?
A. It might radiate harmonics.
B. It radiates the harmonics and fundamental equally well.
C. It is too sharply directional at lower frequencies.
D. It must be neutralized.

Trap antennas have one big disadvantage – on older equipment, they can radiate harmonics. Since ham bands may be harmonically related, it's easy to see that the second harmonic of 7 MHz can radiate from a trap antenna quite nicely because the antenna also has perfect resonance on 14 MHz. So watch out for *harmonics on a trap antenna*. **ANSWER A.**

E9B08 How can the approximate beamwidth in a given plane of a directional antenna be determined?
A. Note the two points where the signal strength of the antenna is 3 dB less than maximum and compute the angular difference.
B. Measure the ratio of the signal strengths of the radiated power lobes from the front and rear of the antenna.
C. Draw two imaginary lines through the ends of the elements and measure the angle between the lines.
D. Measure the ratio of the signal strengths of the radiated power lobes from the front and side of the antenna.

On directional antennas, beamwidth is an important consideration. *Beamwidth is always described by the angle between the two points where the signal strength is down 3 dB from the maximum signal point.* The "tighter" your beam pattern, the smaller the beamwidth. **ANSWER A.**

E9B09 What type of computer program technique is commonly used for modeling antennas?
A. Graphical analysis.
B. Method of Moments.
C. Mutual impedance analysis.
D. Calculus differentiation with respect to physical properties.

The American Radio Relay League has a computer program called *"Method of Moments,"* and this is a great way *for modeling a proposed antenna* on the computer screen. **ANSWER B.**

E9B10 What is the principle of a Method of Moments analysis?
A. A wire is modeled as a series of segments, each having a uniform value of current.
B. A wire is modeled as a single sine-wave current generator.
C. A wire is modeled as a series of points, each having a distinct location in space.
D. A wire is modeled as a series of segments, each having a distinct value of voltage across it.

Using *Method of Moments* (MOM) software, a wire is modeled as a series of segments, *each segment having a distinct value of antenna current*. **ANSWER A.**

E9B13 What does the abbreviation NEC stand for when applied to antenna modeling programs?
A. Next Element Comparison.
B. Numerical Electromagnetics Code.
C. National Electrical Code.
D. Numeric Electrical Computation.

NEC-based antenna modeling stands for Numerical Electromagnetics Code. NEC simulates the electromagnetic response of antennas and metal structures. Jerry Burke and A. Poggio wrote the NEC/MOM family of programs at Lawrence Livermore Labs in 1981, under contract to the U.S. Navy. NEC2 was later released to the public and is now available on most computing platforms. To learn more about NEC, ☞ **Visit: www.nec2.org**. **ANSWER B.**

E9B11 What is a disadvantage of decreasing the number of wire segments in an antenna model below the guideline of 10 segments per half-wavelength?
A. Ground conductivity will not be accurately modeled.
B. The resulting design will favor radiation of harmonic energy.
C. The computed feed point impedance may be incorrect.
D. The antenna will become mechanically unstable.

When modeling antennas on a computer, we always stay above 10 segments per half wavelength so we don't cause the *feed point impedance to be miscalculated using less than* the recommended *10 segments per half wavelength*. **ANSWER C.**

E9B14 What type of information can be obtained by submitting the details of a proposed new antenna to a modeling program?
A. SWR vs. frequency charts.
B. Polar plots of the far-field elevation and azimuth patterns.
C. Antenna gain.
D. All of these choices are correct.

Modern *antenna programs will show SWR* plots at a certain frequency, *polar plots, antenna gain, feed point impedance*, and you can even tweak your proposed antenna by adding in the diameter of those cool-looking polished copper tube elements! **ANSWER D.**

E9B01 In the antenna radiation pattern shown in Figure E9-1, what is the 3-dB beamwidth?

A. 75 degrees.
B. 50 degrees.
C. 25 degrees.
D. 30 degrees.

Looking carefully at the diagram in Figure E9-1, notice that the main lobe of the pattern intersects the 3 dB points at approximately plus and minus 25 degrees. Hence, the *3-dB beamwidth is approximately 50 degrees.* **ANSWER B.**

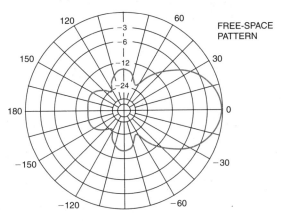

Figure E9-1

E9B02 In the antenna radiation pattern shown in Figure E9-1, what is the front-to-back ratio?

A. 36 dB.
B. 18 dB.
C. 24 dB.
D. 14 dB.

If you look carefully at Figure E9-1, *the rear lobe is about -18 dB down from the front lobe.* **ANSWER B.**

E9B03 In the antenna radiation pattern shown in Figure E9-1, what is the front-to-side ratio?

A. 12 dB.
B. 14 dB.
C. 18 dB.
D. 24 dB.

Study Figure E9-1 carefully, and go with 14 dB as the *front-to-side ratio.* Looking at the figure, this is not an exact measurement, so just stick *14 dB* in your brain before the test. Figure E9-1 will look exactly like it does here in the book when your VE team gives it to you as part of your test papers at your upcoming Extra Class exam. **ANSWER B.**

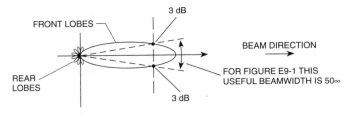

Directional Radiation Pattern of a Yagi Beam

Source: *Antennas – Selection and Installation*, A.J. Evans, Copyright ©1986 Master Publishing, Inc., Niles, Illinois

E9A08 What is meant by antenna gain?

A. The ratio relating the radiated signal strength of an antenna in the direction of maximum radiation to that of a reference antenna.
B. The ratio of the signal in the forward direction to that in the opposite direction.
C. The ratio of the amount of power radiated by an antenna compared to the transmitter output power.
D. The final amplifier gain minus the transmission-line losses, including any phasing lines present.

We normally rate *antenna gain* figures in dB. This is the *numerical ratio relating the radiated signal strength of your antenna to that of another antenna* down the street. Generally, the ham with the higher gain antenna will always get better signal reports. **ANSWER A.**

E9B12 What is the far-field of an antenna?

A. The region of the ionosphere where radiated power is not refracted.
B. The region where radiated power dissipates over a specified time period.
C. The region where radiated field strengths are obstructed by objects of reflection.
D. The region where the shape of the antenna pattern is independent of distance.

The *far field zone* of an antenna is where the *antenna pattern remains constant, independent of distance*. This is generally a safe area when calculating MPE. **ANSWER D.**

E9B04 What may occur when a directional antenna is operated at different frequencies within the band for which it was designed?

A. Feed point impedance may become negative.
B. The E-field and H-field patterns may reverse.
C. Element spacing limits could be exceeded.
D. The gain may change depending on frequency.

Building your own homebrew Yagi can teach some fundamental Yagi antenna concepts:
- Close element spacing will yield narrow bandwidth response.
- Directivity sharpness depends on parasitic-element tuning for maximum gain.
- Boom length, not number of directors, influences gain the most.

A Yagi antenna with closely-spaced reflector and director elements may indeed maximize gain, but *overall gain may have significant variations* if your Yagi design is tuned to the bottom of the band and you begin to operate SSB at the top of the band, several hundred kHz away from your homebrew design. **ANSWER D.**

E9B05 What usually occurs if a Yagi antenna is designed solely for maximum forward gain?

A. The front-to-back ratio increases.
B. The front-to-back ratio decreases.
C. The frequency response is widened over the whole frequency band.
D. The SWR is reduced.

On that brand new Yagi antenna you are computer modeling, the term "front to back ratio" is the ratio of the power radiated in the main lobe to the power being radiated in the opposite direction. As you work your computer antenna modeling program, you will quickly discover that *maximum forward gain will decrease the front to back ratio*. This is why some antenna experts will sacrifice a half dB of forward gain to improve one or two nulls off the back of the beam. **ANSWER B.**

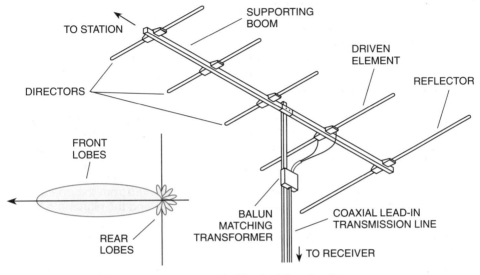

a. Directional Pattern   b. Physical Construction

A beam antenna - the Yagi antenna
Source: *Antennas - Selection and Installation,* © 1986 Master Publishing, Inc., Niles, IL

E9B06 If the boom of a Yagi antenna is lengthened and the elements are properly retuned, what usually occurs?

A. The gain increases.
B. The SWR decreases.
C. The front-to-back ratio increases.
D. The gain bandwidth decreases rapidly.

Longer boom, more gain! **ANSWER A.**

E9B07 How does the total amount of radiation emitted by a directional gain antenna compare with the total amount of radiation emitted from an isotropic antenna, assuming each is driven by the same amount of power?

A. The total amount of radiation from the directional antenna is increased by the gain of the antenna.
B. The total amount of radiation from the directional antenna is stronger by its front to back ratio.
C. They are the same.
D. The radiation from the isotropic antenna is 2.15 dB stronger than that from the directional antenna .

This is a good question to get you thinking about the isotropic antenna – an antenna that radiates equally well in all directions. If you were to pump 100 watts into this theoretical antenna, you would get 100 watts out, unity gain, in every direction, including up and down. Now you switch to the beam antenna and presto, those 100 watts are focused in one direction, with minimum up and minimum

---

down radiation. By taking energy in all directions and concentrating it in just one direction with a beam, you have improved your signal on transmit and receive. But the total amount of radiation emitted is that same 100 watts at the feed point, so there is no magical new power created in the beam, and thus *there is no difference between the two antennas*. A *100 watt* light bulb hanging by a wire and a 100 watt bulb with a polished reflector *is still 100 watts*, right? Sure, but one bulb looks brighter, with the same amount of power, thanks to the reflector! **ANSWER C.**

E3C07 How does the radiation pattern of a horizontally polarized 3-element beam antenna vary with its height above ground?
    A. The main lobe takeoff angle increases with increasing height.
    B. The main lobe takeoff angle decreases with increasing height.
    C. The horizontal beam width increases with height.
    D. The horizontal beam width decreases with height.

When you put up a 3-element Yagi antenna, the high frequency Yagi is usually mounted horizontally and the antenna should be mounted at least one-half wavelength above the ground on the lowest band of operation. Most 3-element Yagis feature tri-band operation for 10 meters, 15 meters and 20 meters. 20-meter band operation would like to see the antenna at a half wavelength off the earth, or about 32 feet in the air. *Anything lower than a half wavelength* from the operating frequency *dramatically raises the main lobe takeoff angle*, and the higher the takeoff angle, the shorter the skip distance. You want to get your beam up at least one-half wavelength above the earth for a nice low takeoff angle for long-range DX. **ANSWER B.**

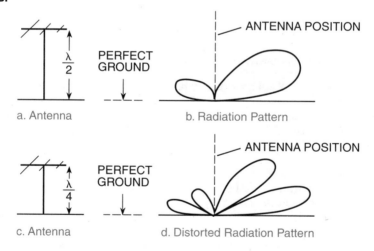

Radiation of 3-Element Yagi for Different Antenna Heights

E3C10 How does the performance of a horizontally polarized antenna mounted on the side of a hill compare with the same antenna mounted on flat ground?
    A. The main lobe takeoff angle increases in the downhill direction.
    B. The main lobe takeoff angle decreases in the downhill direction.
    C. The horizontal beam width decreases in the downhill direction.
    D. The horizontal beam width increases in the uphill direction.

If you live on the East coast, build your home and radio shack on a hill that slopes down to the ocean. Your beam will now have a main lobe whose takeoff angle decreases toward the ocean, giving you a dynamite signal to Europe. Here on the West coast, I am up on some headlands, with the ocean just down the way, giving me a good shot to Asia with the *lower takeoff angle in the downhill direction*. **ANSWER B.**

E9C07 What type of antenna pattern over real ground is shown in Figure E9-2?

A. Elevation.
B. Azimuth.
C. Radiation resistance.
D. Polarization.

Figure E9-2 shows *the takeoff elevation* patterns of an antenna over a real ground source. **ANSWER A.**

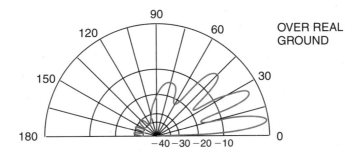

Figure E9-2

E9C08 What is the elevation angle of peak response in the antenna radiation pattern shown in Figure E9-2?

A. 45 degrees.
B. 75 degrees.
C. 7.5 degrees.
D. 25 degrees.

The main zero dB lobe is about 7.5 degrees up, with two other significant lobes at higher elevations. **ANSWER C.**

E9C09 What is the front-to-back ratio of the radiation pattern shown in Figure E9-2?

A. 15 dB.
B. 28 dB.
C. 3 dB.
D. 24 dB.

The antenna *front-to-back ratio* in Figure E9-2 looks to be *about 28 to 30 dB*. This is the ratio of forward "front" radiation to the slight amount of unwanted radiation in the reverse direction from the back of the antenna. This ratio is the same for transmit as well as receive. **ANSWER B.**

E9C10 How many elevation lobes appear in the forward direction of the antenna radiation pattern shown in Figure E9-2?

A. 4.
B. 3.
C. 1.
D. 7.

There are three big lobes, and one smaller lobe in the forward direction. **ANSWER A.**

E9D02 How can linearly polarized Yagi antennas be used to produce circular polarization?

A. Stack two Yagis, fed 90 degrees out of phase, to form an array with the respective elements in parallel planes.

B. Stack two Yagis, fed in phase, to form an array with the respective elements in parallel planes.

C. Arrange two Yagis perpendicular to each other with the driven elements at the same point on the boom and fed 90 degrees out of phase.

D. Arrange two Yagis collinear to each other, with the driven elements fed 180 degrees out of phase.

To get circular polarization out of a pair of Yagis, arrange the *Yagis perpendicular to each other*, with the driven elements in the same plane, and *fed 90 degrees out of phase*. **ANSWER C.**

E9C04 Which of the following describes a basic unterminated rhombic antenna?

A. Unidirectional; four-sides, each side one quarter-wavelength long; terminated in a resistance equal to its characteristic impedance.

B. Bidirectional; four-sides, each side one or more wavelengths long; open at the end opposite the transmission line connection.

C. Four-sides; an LC network at each corner except for the transmission connection.

D. Four-sides, each of a different physical length.

The *resonant rhombic antenna* is a *bidirectional* antenna where *each side is equal to one wavelength*. It is *unterminated* (open at one end), with the transmission line connected to the other end. You better have a lot of wire for a resonant rhombic on 80 meters! **ANSWER B.**

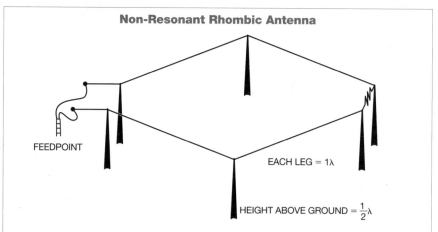

Non-Resonant Rhombic Antenna

FEEDPOINT

EACH LEG = 1λ

HEIGHT ABOVE GROUND = $\frac{1}{2}$λ

The non-resonant rhombic antenna will use a non-inductive resistor to terminate the opposite end from the feedpoint. Feedpoint impedance is typically 600-900 ohms. The terminating non-inductive resistor should be around 800 ohms. The non-resonant rhombic radiates in a substantially uni-directional path toward the terminating resistor. But like its brother, the non-resonant rhombic requires plenty of real estate plus multiple poles to keep each leg absolutely straight.

E9C05 What are the disadvantages of a terminated rhombic antenna for the HF bands?

A. The antenna has a very narrow operating bandwidth.
B. The antenna produces a circularly polarized signal.
C. The antenna requires a large physical area and 4 separate supports.
D. The antenna is more sensitive to man-made static than any other type.

The *big disadvantage* for the terminated *rhombic antenna* for high frequency is the *very large area it will occupy, and the necessity for a minimum of 4 sturdy supports* for proper installation. You better have plenty of real estate! **ANSWER C.**

E9C06 What is the effect of a terminating resistor on a rhombic antenna?

A. It reflects the standing waves on the antenna elements back to the transmitter.
B. It changes the radiation pattern from bidirectional to unidirectional.
C. It changes the radiation pattern from horizontal to vertical polarization.
D. It decreases the ground loss.

A *terminating resistor* at the opposite end of the rhombic from the feed line *changes the radiation pattern from bidirectional to unidirectional.* **ANSWER B.**

Decibels

It is important to know about decibels because they are used extensively in electronics. Look at the derivation for decibels and note it is a measure of the ratio of two powers, P_1 and P_2. Remember the power changes for different dB values. Also, since 6 dB is a four times change, 16 dB (6 dB + 10 dB) will be a 4 × 10 = 40 times change, and 26 dB (6 dB + 20 dB) will be a 4 × 100 = 400 times change. Thus, any dB value from 10 dB and above can be evaluted by using the power change of 1 dB to 9 dB and multiplying it by the 10 dB, 20 dB, 30 dB, 40 dB, 50 dB, 60 dB, etc. change. Thus, 57 dB (7 dB + 50 dB) will be a 5 × 100,000 = 500,000 times change.

dB	Power Change $\dfrac{P_1}{P_2}$		dB	Power Change $\dfrac{P_1}{P_2}$	
1 dB	1 ¼X	Power change	8 dB	6 ¼X	Power change
2 dB	1 ½X	Power change	9 dB	8X	Power change
3 dB	2X	Power change	10 dB	10X	Power change
4 dB	2 ½X	Power change	20 dB	100X	Power change
5 dB	3X	Power change	30 dB	1000X	Power change
6 dB	4X	Power change	40 dB	10,000X	Power change
7 dB	5X	Power change	50 dB	100,000X	Power change
			60 dB	1,000,000X	Power change

Derivation:

$$\text{If dB} = 10 \log_{10} \frac{P_1}{P_2}$$

then what power ratio is 60 dB?

$$60 = 10 \log_{10} \frac{P_1}{P_2}$$

$$\frac{60}{10} = \log_{10} \frac{P_1}{P_2}$$

$$6 = \log_{10} \frac{P_1}{P_2}$$

Remember: logarithm of a number is the exponent to which the base must be raised to get the number.

$$\therefore 10^6 = \frac{P_1}{P_2}$$

$$1,000,000 = \frac{P_1}{P_2}$$

Or $P_1 = 1,000,000 \, P_2$

60 dB means P_1 is one million times P_2

E9D01 How does the gain of an ideal parabolic dish antenna change when the operating frequency is doubled?

A. Gain does not change.
B. Gain is multiplied by 0.707.
C. Gain increases by 6 dB.
D. Gain increases by 3 dB.

From 1270 MHz on up, you may wish to *use a parabolic dish antenna* for microwave work. If you *double the frequency, the gain of the dish increases by 6 dB, a 4 times increase.* **ANSWER C.**

Gordo's Parabolic Antenna for 10GHz Tropo

E9D03 How does the beamwidth of an antenna vary as the gain is increased?

A. It increases geometrically.
B. It increases arithmetically.
C. It is essentially unaffected.
D. It decreases.

The parabolic dish concentrates that beamwidth as gain increases. The higher the gain of a dish, the narrower the beam width. *Beamwidth decreases as gain increases.* **ANSWER D.**

HERE ARE HELPFUL WEBSITES RELATED TO ANTENNAS

http://dxzone.com/cgi/bin/dir/jump2.cgi?ID=14599

www.qsl.net/ta1dx/amator/practical_dipole_antenna.htm

www.bas.thoen.info/gallery/Stealth-Zepp

www.packetradio.com/windom.htm

www.snars.org/newsletter/2011/cs201109.pdf

http://www.antennasoftware.com.ar/

Feedlines and Safety

E9E01 What system matches a high-impedance transmission line to a lower impedance antenna by connecting the line to the driven element in two places spaced a fraction of a wavelength each side of element center?
   A. The gamma matching system.
   B. The delta matching system.
   C. The omega matching system.
   D. The stub matching system.

The *delta matching system* is a popular way *to change the impedance of the transmission line to the impedance of the antenna input.* **ANSWER B.**

E9E02 What is the name of an antenna matching system that matches an unbalanced feed line to an antenna by feeding the driven element both at the center of the element and at a fraction of a wavelength to one side of center?
   A. The gamma match.
   B. The delta match.
   C. The epsilon match.
   D. The stub match.

A *gamma match* is used with coax cable that forms an *unbalanced feed system* going to a tube that acts as a capacitor in series with the connection point. **ANSWER A.**

E9E03 What is the name of the matching system that uses a section of transmission line connected in parallel with the feed line at or near the feed point?
A. The gamma match.
B. The delta match.
C. The omega match.
D. The stub match.

When *a short section of transmission line* is connected to the antenna feed line near the antenna, it is called *stub matching*. Perhaps you used to cut a stub to cancel out interference in your TV receiver. **ANSWER D.**

Impedance Matching Stubs

Quarter wavelength ($\frac{\lambda}{4}$) stubs

(or multiples of odd numbers of $\frac{\lambda}{4}$ s)

Less than $\frac{1}{4}\lambda$ stub

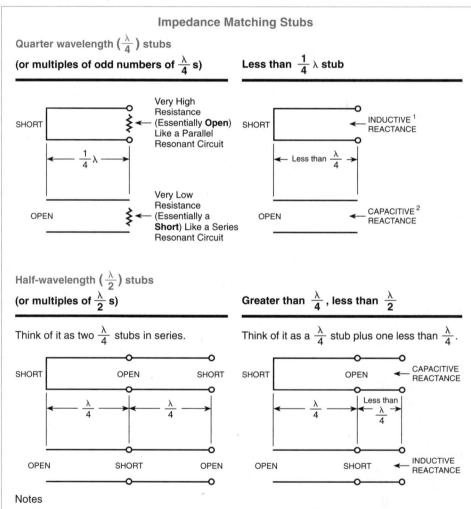

Very High Resistance (Essentially **Open**) Like a Parallel Resonant Circuit

Very Low Resistance (Essentially a **Short**) Like a Series Resonant Circuit

INDUCTIVE [1] REACTANCE

CAPACITIVE [2] REACTANCE

Half-wavelength ($\frac{\lambda}{2}$) stubs

(or multiples of $\frac{\lambda}{2}$ s)

Think of it as two $\frac{\lambda}{4}$ stubs in series.

Greater than $\frac{\lambda}{4}$, less than $\frac{\lambda}{2}$

Think of it as a $\frac{\lambda}{4}$ stub plus one less than $\frac{\lambda}{4}$.

Notes
1. Since a short at end makes the diagram look like a long wire, this should remind the reader that impedance is an inductive reactance.
2. Since an open at end makes the diagram look like two parallel plates, this should remind the reader that impedance is a capacitive reactance.

E9E04 What is the purpose of the series capacitor in a gamma-type antenna matching network?

A. To provide DC isolation between the feed line and the antenna.
B. To cancel the inductive reactance of the matching network.
C. To provide a rejection notch to prevent the radiation of harmonics.
D. To transform the antenna impedance to a higher value.

In some *gamma*-type antenna *matching* networks, the center conductor and PVC insulation on the center conductor of coaxial cable is used inside a tube, as a series capacitance. *Series capacitance will cancel the inductive reactance* of the matching network, leading to resonance. **ANSWER B.**

E9E05 How must the driven element in a 3-element Yagi be tuned to use a hairpin matching system?

A. The driven element reactance must be capacitive.
B. The driven element reactance must be inductive.
C. The driven element resonance must be lower than the operating frequency.
D. The driven element radiation resistance must be higher than the characteristic impedance of the transmission line.

For the *hairpin matching system* to work well, the *driven element reactance must be capacitive, X_C*. **ANSWER A.**

E9E06 What is the equivalent lumped-constant network for a hairpin matching system on a 3-element Yagi?

A. Pi network.
B. Pi-L network.
C. L network.
D. Parallel-resonant tank.

We use an *L-network* to match a 3-element Yagi for a *hairpin matching system*. **ANSWER C.**

E9E07 What term best describes the interactions at the load end of a mismatched transmission line?

A. Characteristic impedance.
B. Reflection coefficient.
C. Velocity factor.
D. Dielectric constant.

A *mismatched* transmission line will give you *reflections*. **ANSWER B.**

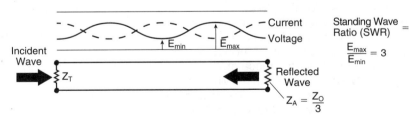

Impedance Mismatch Causes Reflected Wave

E9E08 Which of the following measurements is characteristic of a mismatched transmission line?
    A. An SWR less than 1:1.
    B. A reflection coefficient greater than 1.
    C. A dielectric constant greater than 1.
    D. An SWR greater than 1:1.

If the *SWR is greater than 1:1*, this describes a *mismatched* transmission line. **ANSWER D.**

SWR Bridge

E9E09 Which of these matching systems is an effective method of connecting a 50-ohm coaxial cable feed line to a grounded tower so it can be used as a vertical antenna?
    A. Double-bazooka match.   C. Gamma match.
    B. Hairpin match.   D. All of these choices are correct.

I have a grounded tower that I can use on 160 meters, running the cable about halfway up, into a *gamma match, that turns the tower into a halfwave antenna!* The outer braid taps the metal tower leg and I run the center conductor, extended, parallel to the same tower leg, and then making a sharp turn and connect it to the same leg, allowing for the impedance to match the transmission line. If you were to look at it with an ohm meter, it would look like a direct short – but if you work from an antenna book, your gamma match and a manual tuner can get you down to the band of choice. **ANSWER C.**

E9E10 Which of these choices is an effective way to match an antenna with a 100-ohm feed point impedance to a 50-ohm coaxial cable feed line?
    A. Connect a 1/4-wavelength open stub of 300-ohm twin-lead in parallel with the coaxial feed line where it connects to the antenna.
    B. Insert a 1/2 wavelength piece of 300-ohm twin-lead in series between the antenna terminals and the 50-ohm feed cable.
    C. Insert a 1/4-wavelength piece of 75-ohm coaxial cable transmission line in series between the antenna terminals and the 50-ohm feed cable.
    D. Connect 1/2 wavelength shorted stub of 75-ohm cable in parallel with the 50-ohm cable where it attaches to the antenna.

In VHF and UHF antenna stacking, we many times end up with the combined antenna system presenting an impedance of 100 ohms to feed the coax, rather than 50 ohms. *Calculate one quarter wavelength*, including coax velocity factor for your 75 ohm matching network (usually 0.66), *and insert it between the antenna terminals and the 50 ohm coax leading down to the ham shack.* **ANSWER C.**

E9E11 What is an effective way of matching a feed line to a VHF or UHF antenna when the impedances of both the antenna and feed line are unknown?

    A. Use a 50-ohm 1:1 balun between the antenna and feed line.

    B. Use the "universal stub" matching technique.

    C. Connect a series-resonant LC network across the antenna feed terminals.

    D. Connect a parallel-resonant LC network across the antenna feed terminals.

The *universal stub* takes unwanted reactance out of the driven portion of the antenna system, without actually computing impedances up the tower. You'll need to use your trusty portable SWR analyzer as you adjust the stiff bare wire shorting elements normally connected to the driven element. **ANSWER B.**

E9E12 What is the primary purpose of a phasing line when used with an antenna having multiple driven elements?

    A. It ensures that each driven element operates in concert with the others to create the desired antenna pattern.

    B. It prevents reflected power from traveling back down the feed line and causing harmonic radiation from the transmitter.

    C. It allows single-band antennas to operate on other bands.

    D. It makes sure the antenna has a low-angle radiation pattern .

As an Extra Class operator, you likely will choose an exotic, multiband beam antenna to exercise your new privileges. Multiband antennas have a variety of methods to phase most elements for the desired beam pattern. "Trapless" designs may use rigid *phasing lines* to *ensure each driven element operates in concert with all of the other elements* on the antenna for the desired antenna pattern. **ANSWER A.**

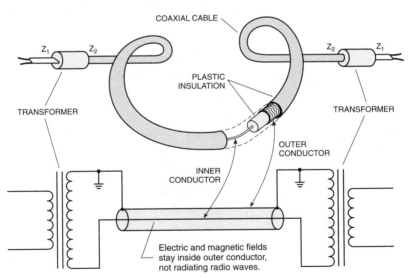

Coaxial Cable

E9E13 What is the purpose of a Wilkinson divider?
A. It divides the operating frequency of a transmitter signal so it can be used on a lower frequency band.
B. It is used to feed high-impedance antennas from a low-impedance source.
C. It divides power equally among multiple loads while preventing changes in one load from disturbing power flow to the others.
D. It is used to feed low-impedance loads from a high-impedance source.
With many high frequency antennas now incorporating multiple-band, driven elements, all off one feed line, the *Wilkinson divider distributes the antenna input power equally* to the multiple driven elements *while preventing one driven element from disturbing another.* **ANSWER C.**

E9F08 What is the term for the ratio of the actual speed at which a signal travels through a transmission line to the speed of light in a vacuum?
A. Velocity factor.   C. Surge impedance.
B. Characteristic impedance.   D. Standing wave ratio.
Velocity factor is a ratio of how much slower radio signals travel in a conductor compared to the velocity of radio waves in free space. Radio waves in free space travel at the speed of light. In coaxial cable, and along antenna conductors, the radio waves travel slower. **ANSWER A.**

E9F01 What is the velocity factor of a transmission line?
A. The ratio of the characteristic impedance of the line to the terminating impedance.
B. The index of shielding for coaxial cable.
C. The velocity of the wave in the transmission line multiplied by the velocity of light in a vacuum.
D. The velocity of the wave in the transmission line divided by the velocity of light in a vacuum.
In free space, radio waves travel at 300 million meters per second. But in coaxial cable transmission lines, and other types of feed lines, the radio waves move slower. *Velocity factor* is the *velocity of the radio wave on the transmission line divided by the velocity in free space.* **ANSWER D.**

E9F02 Which of the following determines the velocity factor of a transmission line?
A. The termination impedance.   C. Dielectric materials used in the line.
B. The line length.   D. The center conductor resistivity.
There are some great low-loss coaxial cable types out there in radio land with different dielectrics separating the center conductor and the outside shield. It is the *type of dielectric used that determines the velocity factor.* **ANSWER C.**

E9F04 What is the typical velocity factor for a coaxial cable with solid polyethylene dielectric?
A. 2.70.   C. 0.30.
B. 0.66.   D. 0.10.
Most coax cable has a velocity factor of 0.66. This is not stated on the outside jacket, and may not even be known by the dealer selling the coax. You can contact the manufacturer or distributor directly to find out the actual velocity factor. **ANSWER B.**

E9F03 Why is the physical length of a coaxial cable transmission line shorter than its electrical length?

A. Skin effect is less pronounced in the coaxial cable.
B. The characteristic impedance is higher in a parallel feed line.
C. The surge impedance is higher in a parallel feed line.
D. Electrical signals move more slowly in a coaxial cable than in air.

Radio frequency energy moves slightly *slower inside coaxial cable* than it does in free space. **ANSWER D.**

E9F05 What is the approximate physical length of a solid polyethylene dielectric coaxial transmission line that is electrically one-quarter wavelength long at 14.1 MHz?

A. 20 meters.
B. 2.3 meters.
C. 3.5 meters.
D. 0.2 meters.

When you work on antenna phasing harnesses, you will need to calculate the velocity factor for a piece of coax in order to know where to place the connectors. You must be accurate down to a fraction of an inch! You can calculate the physical length of a coax cable which is electrically one-quarter wavelength long using the formula:

$$L \text{ (in feet)} = \frac{984\lambda V}{f}$$

Where: L is antenna length in **feet**
λ is wavelength in **meters**
V is **velocity factor**
f is frequency in **MHz**

For this problem, multiply 984 × 0.25, because the coax is one quarter wavelength. Then multiply the result by 0.66, the velocity factor. Remember, 0.66 is the velocity factor for coax. We are assuming that value for these problems. Divide the total by 14.1, the frequency in MHz. The length comes out to about 11.51 feet. The answer needs to be in meters, so divide 11.5 by 3 (there are about 3 feet to a meter). The result is 3.83, but the *3.5 meters* in answer C is the closest. (To calculate coax in meters, simply substitute 300 for 984. This will eliminate one step in the conversion!) **ANSWER C.**

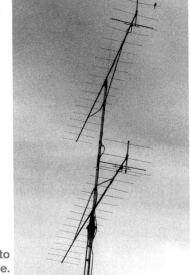

Proper coax cable length is important to maximize phased antenna performance.

E9F06 What is the approximate physical length of an air-insulated, parallel conductor transmission line that is electrically one-half wavelength long at 14.10 MHz?

A. 15 meters.

B. 20 meters.

C. 10 meters.

D. 71 meters.

In this problem, the parallel conductor feed line one-half wavelength long at 14.10 MHz, with a velocity factor of 0.95, is calculated by multiplying *984 × 0.5 × 0.95* and *dividing by 14.10*, the frequency in MHz. *Convert feet to meters*, and you end up with *10 meters.* **ANSWER C.**

E9F07 How does ladder line compare to small-diameter coaxial cable such as RG-58 at 50 MHz?

A. Lower loss.

B. Higher SWR.

C. Smaller reflection coefficient.

D. Lower velocity factor.

Going to 450-ohm ladder line on the 6-meter band is one way to reduce losses over typical RG-58 coax cable, and keep your expenses to a minimum. They both cost about the same. RG-58 should be relegated to test leads, never used for antenna feed line runs. The loss is simply too high. *450 ohm ladder line will have much lower loss at 50 MHz than small RG-58 coax.* **ANSWER A.**

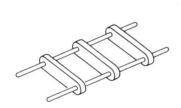

Outer Insulation

Copper Braid Shield

Inner Conductor

Polyethylene Dielectric

Ladder Line

Coax

E9F09 What is the approximate physical length of a solid polyethylene dielectric coaxial transmission line that is electrically one-quarter wavelength long at 7.2 MHz?

A. 10 meters.

B. 6.9 meters.

C. 24 meters.

D. 50 meters.

The coax cable is again one-quarter wavelength, so multiple 984 x 0.25, and the result by 0.66, the velocity factor. We again will assume this value. You must know it. Now divide these three multiplied figures by the frequency, 7.2 MHz, and you end up with 22.55 feet of coax. Dividing by 3 to get approximate length of 7.5 meters, leads you to *6.9 meters*. (See formula at E9F05) **ANSWER B.**

E9F10 What impedance does a 1/8-wavelength transmission line present to a generator when the line is shorted at the far end?

A. A capacitive reactance.

B. The same as the characteristic impedance of the line.

C. An inductive reactance.

D. The same as the input impedance to the final generator stage.

If a *1/8 wavelength transmission line is shorted out at the far end*, it will look like an *inductive reactance at the generator*. **ANSWER C.**

E9F11 What impedance does a 1/8-wavelength transmission line present to a generator when the line is open at the far end?
  A. The same as the characteristic impedance of the line.
  B. An inductive reactance.
  C. A capacitive reactance.
  D. The same as the input impedance of the final generator stage.
If that *1/8 wavelength transmission line is open* at the far end, it will look like a *capacitive reactance* at the generator. **ANSWER C.**

E9F12 What impedance does a 1/4-wavelength transmission line present to a generator when the line is open at the far end?
  A. The same as the characteristic impedance of the line.
  B. The same as the input impedance to the generator.
  C. Very high impedance.
  D. Very low impedance .
If the far end of a *quarter-wavelength transmission line is open*, it will look like a *very low impedance* at the generator. **ANSWER D.**

E9F13 What impedance does a 1/4-wavelength transmission line present to a generator when the line is shorted at the far end?
  A. Very high impedance.
  B. Very low impedance.
  C. The same as the characteristic impedance of the transmission line.
  D. The same as the generator output impedance.
On a *quarter-wavelength transmission line, shorting* it at the far end will look like a *very high impedance* at the generator. **ANSWER A.**

E9F14 What impedance does a 1/2-wavelength transmission line present to a generator when the line is shorted at the far end?
  A. Very high impedance.
  B. Very low impedance.
  C. The same as the characteristic impedance of the line.
  D. The same as the output impedance of the generator.
If a half-wavelength transmission *line is shorted* at the far end, it will have a *very low impedance*, essentially a short. **ANSWER B.**

E9F15 What impedance does a 1/2-wavelength transmission line present to a generator when the line is open at the far end?
  A. Very high impedance.
  B. Very low impedance.
  C. The same as the characteristic impedance of the line.
  D. The same as the output impedance of the generator.
If that *half-wavelength transmission line is open* at the far end, it will have a *very high impedance*, essentially an open circuit. **ANSWER A.**

Why is Characteristic Impedance So Important?

Let's look at this Figure to help us understand why the characteristic impedance, Z_0, of a transmission line is so important for the efficient transfer of energy. Figure a shows a transmitter coupled to an antenna with a transmission line that has a characteristic impedance of Z_0. The transmitter's output impedance is Z_T and the antenna inputimpedance is Z_A, the load impedance that the antenna presents to the transmission line.

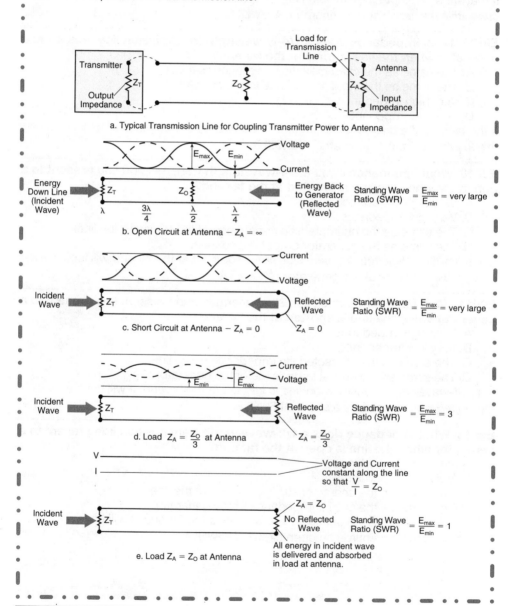

a. Typical Transmission Line for Coupling Transmitter Power to Antenna

Energy Down Line (Incident Wave)

Energy Back to Generator (Reflected Wave)

Standing Wave Ratio (SWR) $= \dfrac{E_{max}}{E_{min}} =$ very large

b. Open Circuit at Antenna $- Z_A = \infty$

Incident Wave

Reflected Wave

Standing Wave Ratio (SWR) $= \dfrac{E_{max}}{E_{min}} =$ very large

c. Short Circuit at Antenna $- Z_A = 0$

Incident Wave

Reflected Wave

Standing Wave Ratio (SWR) $= \dfrac{E_{max}}{E_{min}} = 3$

d. Load $Z_A = \dfrac{Z_0}{3}$ at Antenna

Voltage and Current constant along the line so that $\dfrac{V}{I} = Z_0$

Incident Wave

No Reflected Wave

Standing Wave Ratio (SWR) $= \dfrac{E_{max}}{E_{min}} = 1$

All energy in incident wave is delivered and absorbed in load at antenna.

e. Load $Z_A = Z_0$ at Antenna

Let's assume that Z_A is infinity — in other words, the load on the transmission line at the antenna is an open circuit. Figure b is a diagram of the interconnection and shows the voltages and currents that exist on the transmission line. Since Z_A is an open circuit, the current at the end of the line is zero and the voltage is maximum. Since the wave of energy sent from the transmitter has no load to absorb the energy, it is reflected back to the transmitter. The voltage and current waveforms shown are the combined resultant of the wave from the transmitter (called the incident wave) and the reflected wave from the load end. In this case, the total wave is reflected back and no signal is transmitted.

Standing-Wave Ratio

The ratio of the maximum amplitude, E_{max}, of the resultant voltage waveform on the line to the minimum amplitude, E_{min}, is called the Standing-Wave Ratio, SWR, or more correctly the Voltage Standing-Wave Ratio, VSWR. In the case of an open load, since there is no energy absorbed at the load and all energy delivered is absorbed as losses in the line or absorbed by the transmitter's output resistance, the VSWR is very high (in the order of 20 or greater). VSWR is a measure of how well energy is being transferred over a transmission line to the load. VSWRs below 1.5 indicate quite efficient transfer of energy (VSWR = 1 is perfect), and VSWRs above 2 begin to indicate poor matching.

Now let's look at Figure c. Here the Z_A is a short circuit. That means the voltage is zero at the end of the line and the current is maximum. Again, no energy is absorbed by the load. It is all reflected back down the line. The voltage and current waveforms vary in minimums and maximums every quarter wavelength, as in the open-circuit case, but are 180° out of phase from the open-circuit case. The standing-wave ratio again is very large and no energy is transferred.

A Matched Case — ZA = Z0

Figure e is the case where all the energy in the incident wave is absorbed by the load. There is no reflected wave because $Z_A = Z_0$. The antenna load is matched to the characteristic impedance of the transmission line and maximum transfer of power from the transmitter to the antenna occurs. The standing-wave ratio is 1 because the voltage and current in the line are constant and are such that they equal Z_0 everywhere along the line. Thus, we see how important it is to match the load impedance to the transmission line Z_0. With $Z_A = Z_0$, there will be some small losses in the transmission line, but if there is a mismatch, energy will be reflected from the load back to the transmitter and large losses on the line will be dissipated as heat.

Source: *Basic Communications Electronics*, Hudson & Luecke, © 1999 Master Publishing, Inc., Niles, IL

E9F16 Which of the following is a significant difference between foam-dielectric coaxial cable and solid-dielectric cable, assuming all other parameters are the same?
A. Reduced safe operating voltage limits.
B. Reduced losses per unit of length.
C. Higher velocity factor.
D. All of these choices are correct.

When wiring your new ham station's antenna system, go with the best coaxial cable money can buy. On high frequency bands, coax cable losses are significantly less than operating on VHF and UHF. Popular RG-8 foam coax, on the 15 meter band, represents only 0.7 db loss per 100 feet. Foam coax is lighter weight than solid dielectric cable, but foam dielectric reduces safe operating voltage limits and has a slightly-higher velocity factor than solid dielectric. Recently, land mobile rated (LMR) coax cables, about the same size as RG-8, exhibit wonderful qualities of being lightweight, have a non-contaminating jacket to resist sunlight cracks, exhibit a common 0.66 velocity factor, and have slightly less attenuation on high frequency bands. It handles a little bit like a cold garden hose but, nonetheless, LMR type coax has plenty of benefits. For this question, *all of the answer choices are correct for the difference between foam dielectric coax and solid-dielectric coax.* **ANSWER D.**

Coax Cable Type, Size and Loss per 100 Feet			
Coax Type	Size	Loss at HF 100 MHz	Loss at UHF 400 MHz
RG-6	Large	2.3 dB	4.7 dB
RG-59	Medium	2.9 dB	5.9 dB
RG-58U	Small	4.3 dB	9.4 dB
RG-8X	Medium	3.7 dB	8.0 dB
RG-8U	Large	1.9 dB	4.1 dB
RG-213	Large	1.9 dB	4.5 dB
Hardline	Large, Rigid	0.5 dB	1.5 dB

E4A07 Which of the following is an advantage of using an antenna analyzer compared to an SWR bridge to measure antenna SWR?
A. Antenna analyzers automatically tune your antenna for resonance.
B. Antenna analyzers do not need an external RF source.
C. Antenna analyzers display a time-varying representation of the modulation envelope.
D. All of these choices are correct.

Likely no other ham radio accessory product has offered ham radio experimenters a bigger breakthrough in measuring antenna and feed line SWR than the *antenna analyzer.* Credit goes to MFJ Corporation for their antenna analyzer – an inexpensive, battery-operated micro VFO transmitter that also has a built-in SWR bridge capable of spanning from 2 MHz through UHF. This allows you to check the antenna feed point impedance and resonance up on the roof, rather than going back to the ham shack and putting a big test signal out on the air. Since the *micro transmitter is built in to the analyzer, you don't need an actual radio transmitter to test your antenna system.* **ANSWER B.**

SWR Analyzer

Courtesy of MFJ Enterprises, Inc.

E4B11 How should a portable antenna analyzer be connected when measuring antenna resonance and feed point impedance?

A. Loosely couple the analyzer near the antenna base.
B. Connect the analyzer via a high-impedance transformer to the antenna.
C. Connect the antenna and a dummy load to the analyzer.
D. Connect the antenna feed line directly to the analyzer's connector.

When I am working on a rooftop-mounted, high-frequency antenna, I carry my trusty, portable SWR antenna analyzer and a short coax jumper cable that feeds my analyzer directly to the antenna input. The *SWR antenna analyzer* contains a signal generator that takes the place of the transmitter when testing an antenna or feed line, so it should be *directly connected to the feed line.* **ANSWER D.**

Frequency counter and SWR indicator

Courtesy of MFJ

E9G01 Which of the following can be calculated using a Smith chart?

A. Impedance along transmission lines.
B. Radiation resistance.
C. Antenna radiation pattern.
D. Radio propagation.

The *Smith Chart* is an invaluable graph for *calculating the impedance along Amateur Radio transmission lines.* Many times you will see a completed Smith Chart, generated by a computer, illustrating the resonance of a commercial-quality VHF or UHF collinear base station and repeater antenna. **ANSWER A.**

Smith Chart—Finding Antenna Impedance

Finding Antenna Impedance, Z_A, when Fed with Known Length of Transmission Line (λ in Wavelengths) of Given Impedance, Z_o.
(Impedance and wavelength of transmission line are at the frequency of transmission)

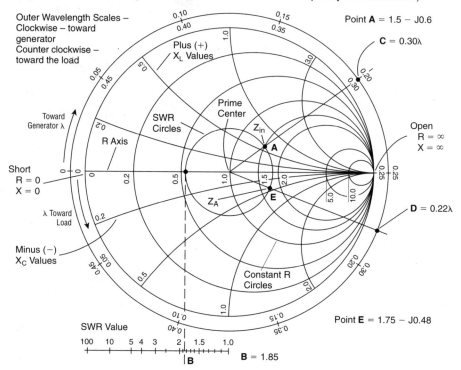

Outer Wavelength Scales –
Clockwise – toward generator
Counter clockwise – toward the load

Point A = 1.5 − J0.6

C = 0.30λ

Plus (+) X_L Values

Prime Center

Z_{in}

SWR Circles

Toward Generator λ

R Axis

Open
R = ∞
X = ∞

Short
R = 0
X = 0

Z_A

λ Toward Load

Minus (−) X_C Values

Constant R Circles

D = 0.22λ

Point E = 1.75 − J0.48

SWR Value

100 10 5 4 3 2 1.5 1.0

B = 1.85

Known Values: $Z_o = 50\Omega$, $\lambda_z = 0.42\lambda$

Steps

1. Connect transmission line to antenna.
2. Measure the input impedance Z_{in} to transmission line. (Example 75 + J30)
3. Normalize Z_{in} by dividing by Z_o.

$$Z_{in} = \frac{75}{50} + J\frac{30}{50} = 1.5 + J0.6$$

4. Plot 1.5 + J0.6 on chart (point A is at intersection of a resistance circle of 1.5 and a positive reactance curve of 0.6)
5. Draw SWR circle with prime center as center and radius through A. Project tangent to SWR circle at point where circle intersects the resistance axis down to linear scale and read value of SWR (Point B = 1.85).
6. Draw a line (radial line) from prime center through Z_{in} at A and project to wavelength scale. Use "toward load" scale because you are going from Z_{in} at generator to the load presented by Antenna. (Point C = 0.30λ)

7. To find the load impedance line on the wavelength (towards load) scale, add the transmission line length to Point C value. (Point D = 0.42 + .30 = 0.72λ)
8. Since the wavelength scales are only plotted for 0.5λ, the values repeat every 0.5λ. Therefore, subtract 0.5 for 0.72 to arrive at 0.22. (Point D = 0.22 toward load)
9. Draw another radial line from prime center to Point D. The intersection with the SWR circle is the load impedance of the antenna. (Point E = 1.75 − J0.48). You may have to interprolate between lines.
10. This is still a normalized value so multiply by $Z_o = 50\Omega$ to obtain true value or antenna impedance. $Z_A = 50$ (1.75 − J0.48) = 87.5 − J24. Which is an impedance of 87.5 ohms of resistance and 24 ohms of capacitive reactance.

E4A08 Which of the following instruments would be best for measuring the SWR of a beam antenna?

A. A spectrum analyzer.
B. A Q meter.
C. An ohmmeter.
D. An antenna analyzer.

I always take my *portable antenna analyzer* when I go out for any ham radio station checkout. If I'm at the top of a sailboat mast, or up a tower, or down inside a boat's lazarette, my portable analyzer checks everything from automatic antenna tuners, marine VHF antennas, and the big 3-element beam up 90 feet in the air. **ANSWER D.**

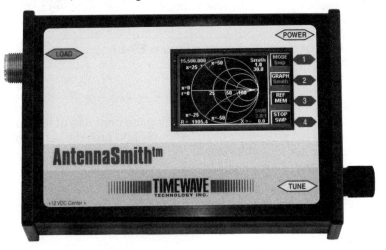

The AntennaSmith is an automatic antenna impedance analyzer. Built-in software allows you to instantly see the effect of an adjustment or a change made to your antenna system. No computer connection is required to see all the graphs – even the Smith chart – and the unit can run on battery power for use in the field.

E9G02 What type of coordinate system is used in a Smith chart?

A. Voltage circles and current arcs.
B. Resistance circles and reactance arcs.
C. Voltage lines and current chords.
D. Resistance lines and reactance chords.

If you look carefully at a *Smith Chart*, you will see *two circles – resistance circles* centered on a resistance axis – all tangent to an open circuit point (infinity resistance and infinity reactance), *and* portions of *reactance arcs* fanning out from the resistance axis and passing through the same open circuit point. Inductive reactance curves are plotted positively from the resistance axis, and capacitive reactance curves negatively. **ANSWER B.**

E9G03 Which of the following is often determined using a Smith chart?

A. Beam headings and radiation patterns.
B. Satellite azimuth and elevation bearings.
C. Impedance and SWR values in transmission lines.
D. Trigonometric functions.

Commercial-quality coaxial cable, coiled on a 500-foot spool, may also contain a *Smith Chart* that has been computer-generated to show *impedance and SWR value*. **ANSWER C.**

E9G04 What are the two families of circles and arcs that make up a Smith chart?
A. Resistance and voltage.
B. Reactance and voltage.
C. Resistance and reactance.
D. Voltage and impedance.

Resistance and reactance values comprise the "two families of circles" on the Smith Chart. Circles for positive inductive reactance are to the right of the resistance axis in the figure below; circles for negative capacitance reactance are to the left of the resistance axis. **ANSWER C.**

E9G05 What type of chart is shown in Figure E9-3?
A. Smith chart.
B. Free-space radiation directivity chart.
C. Elevation angle radiation pattern chart.
D. Azimuth angle radiation pattern chart.

The *Smith chart* is an excellent way of visualizing the contours of resistance or reactance. The chart is used for circuit analysis and provides a graphic presentation for this purpose. **ANSWER A.**

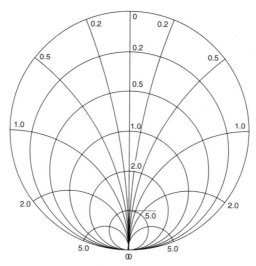

Figure E9-3

E9G06 On the Smith chart shown in Figure E9-3, what is the name for the large outer circle on which the reactance arcs terminate?

A. Prime axis.   C. Impedance axis.
B. Reactance axis.   D. Polar axis.

The large outer circle is the reactance axis. Within the reactance axis are curved reactance circles that are tangent to the resistance axis. Where the reactance circle intersects with the reactance axis, you will see an assigned value of reactance. Inductive reactance is to the right, and capacitive reactance is to the left. **ANSWER B.**

E9G07 On the Smith chart shown in Figure E9-3, what is the only straight line shown?
A. The reactance axis.
B. The current axis.
C. The voltage axis.
D. The resistance axis.

On a *Smith chart*, the *straight line is* always the *resistance axis*. **ANSWER D.**

E9G08 What is the process of normalization with regard to a Smith chart?
A. Reassigning resistance values with regard to the reactance axis.
B. Reassigning reactance values with regard to the resistance axis.
C. Reassigning impedance values with regard to the prime center.
D. Reassigning prime center with regard to the reactance axis.

In order that the Smith chart coordinate system apply to all types of transmission lines, the coordinates are "customized" to a particular transmission line by dividing test measurements by the characteristic impedance of the transmission line. This *process of reassigning resistance and reactance values with regard to the prime center is called "normalizing."* **ANSWER C.**

E9G09 What third family of circles is often added to a Smith chart during the process of solving problems?
A. Standing-wave ratio circles.
B. Antenna-length circles.
C. Coaxial-length circles.
D. Radiation-pattern circles.

When a computer analyzes an antenna design, it will draw a *third family of circles* on the Smith chart *showing standing-wave ratios*. **ANSWER A.**

E9G10 What do the arcs on a Smith chart represent?
A. Frequency.
B. SWR.
C. Points with constant resistance.
D. Points with constant reactance.

The arcs (curved lines) on a Smith chart *are portions of reactance circles*. Remember that the large outer circle bounding the coordinate portion of the chart is the reactance axis. **ANSWER D.**

E9G11 How are the wavelength scales on a Smith chart calibrated?
A. In fractions of transmission line electrical frequency.
B. In fractions of transmission line electrical wavelength.
C. In fractions of antenna electrical wavelength.
D. In fractions of antenna electrical frequency.

The *wavelength scales* on the Smith chart are calibrated in *portions of transmission line electrical wavelength*. **ANSWER B.**

E9H04 What term describes station output, including the transmitter, antenna and everything in between, when considering transmitter power and system gains and losses?
A. Power factor.
B. Half-power bandwidth.
C. Effective radiated power.
D. Apparent power.

It's important to always calculate *effective radiated power* when you set up your ham station. The table at E9F16 shows you the dB's you could lose with cheap coax cable, and the valuable dB's you can gain by using a quality collinear VHF or UHF antenna. **ANSWER C.**

E9H01 What is the effective radiated power relative to a dipole of a repeater station with 150 watts transmitter power output, 2-dB feed line loss, 2.2-dB duplexer loss and 7-dBd antenna gain?

A. 1977 watts.

B. 78.7 watts.

C. 420 watts.

D. 286 watts.

Our total loss is about 4.2 dB, and our total gain is 7 dB. This is an *approximate 3 dB overall gain*, and this almost *doubles your 150 watts to an approximate 286 watts*. **ANSWER D.**

E9H02 What is the effective radiated power relative to a dipole of a repeater station with 200 watts transmitter power output, 4-dB feed line loss, 3.2-dB duplexer loss, 0.8-dB circulator loss and 10-dBd antenna gain?

A. 317 watts.

B. 2000 watts.

C. 126 watts.

D. 300 watts.

Our total loss is 8 dB, and our total gain is 10 dB gain. This gives us a net gain of 2 dB; therefore, from Decibels chart, we have a gain of 1.585 times. Therefore, with 200 watts input, *the ERP is 317 watts (200 × 1.585)*.

To check your answer, remember

$$dB = 10 \log_{10} \frac{P_1}{P_2}$$

where: P_1 = Power Output

P_2 = Power Input

Therefore,

$$dB = 10 \log_{10} \frac{317}{200}$$

$$dB = 10 \log_{10} 1.585 \qquad \text{The } \log_{10} \text{ of 1.585 is 0.2}$$

$$dB = 10 \times 0.2 = 2$$

Since the output power is 317 watts and the input power is 200 watts, there is a net gain of 2 dB. **ANSWER A.**

E9H03 What is the effective isotropic radiated power of a repeater station with 200 watts transmitter power output, 2-dB feed line loss, 2.8-dB duplexer loss, 1.2-dB circulator loss and 7-dBi antenna gain?

A. 159 watts.

B. 252 watts.

C. 632 watts.

D. 63.2 watts.

Here is our last calculation with dB values. Our total loss adds up to a 6 dB loss, and our total gain is 7 dB, for an actual gain of 1 dB. This would be a slight increase to our 200 watts output in effective radiated power, so *ERP is 252 watts*. **ANSWER B.**

E0A04 When evaluating a site with multiple transmitters operating at the same time, the operators and licensees of which transmitters are responsible for mitigating over-exposure situations?

A. Only the most powerful transmitter.

B. Only commercial transmitters.

C. Each transmitter that produces 5% or more of its MPE exposure limit at accessible locations.

D. Each transmitter operating with a duty-cycle greater than 50%.

If you are setting up a new repeater station on a mountaintop with many other transmitters, and *if your station produces 5% or more of the maximum permissible* exposure at that location, it will be *your responsibility for mitigating* this 5% over of the maximum permissible exposure. **ANSWER C.**

E9D04 Why is it desirable for a ground-mounted satellite communications antenna system to be able to move in both azimuth and elevation?

A. In order to track the satellite as it orbits the Earth.
B. So the antenna can be pointed away from interfering signals.
C. So the antenna can be positioned to cancel the effects of Faraday rotation.
D. To rotate antenna polarization to match that of the satellite.

You need both an *azimuth* as well as an *elevation* rotor control *to track the satellites* up in orbit. I hope you will also join AMSAT to support amateur satellites – visit www.amsat.org to learn more. **ANSWER A.**

E0A11 Which of the following injuries can result from using high-power UHF or microwave transmitters?

A. Hearing loss caused by high voltage corona discharge.
B. Blood clotting from the intense magnetic field.
C. Localized heating of the body from RF exposure in excess of the MPE limits.
D. Ingestion of ozone gas from the cooling system.

If you plan to join me on a mountaintop for some 10 GHz microwave activity, stay away from the "business end" of my dish antenna, and watch out for any misaligned wave guides that could leak microwave radiation frequency emissions. *Microwave energy heats the body* with RF exposure in excess of maximum permissible exposure limits. This is what happens inside the microwave oven in your kitchen. Steer clear of any beam of microwave energy coming from the antenna system. **ANSWER C.**

E0A10 What material found in some electronic components such as high-voltage capacitors and transformers is considered toxic?

A. Polychlorinated biphenyls.
B. Polyethylene.
C. Polytetrafluroethylene.
D. Polymorphic silicon.

You know the dangers of power pole transformers that contain *PCBs (PolyChlorinated Biphenyls)*. **ANSWER A.**

E0A09 Which insulating material commonly used as a thermal conductor for some types of electronic devices is extremely toxic if broken or crushed and the particles are accidentally inhaled?

A. Mica.   C. Beryllium oxide.
B. Zinc oxide.   D. Uranium hexafluoride.

Be careful around large antennas that may contain beryllium oxide on the metal internal contactors. These antennas are terrific performers, and are safe, but, if ever you decide to take one apart, remember that *beryllium oxide is extremely toxic if inhaled.* **ANSWER C.**

E0A05 What is one of the potential hazards of using microwaves in the amateur radio bands?

A. Microwaves are ionizing radiation.
B. The high gain antennas commonly used can result in high exposure levels.
C. Microwaves often travel long distances by ionospheric reflection.
D. The extremely high frequency energy can damage the joints of antenna structures.

I run 10 GHz microwave, and a relatively large 20 dB dish antenna, and I am always cautious to never let anyone stand in the antenna main lobe when transmitting. I don't want to *"cook" my neighbors with my microwave transmissions*! **ANSWER B.**

E0A03 Which of the following would be a practical way to estimate whether the RF fields produced by an amateur radio station are within permissible MPE limits?

A. Use a calibrated antenna analyzer.
B. Use a hand calculator plus Smith-chart equations to calculate the fields.
C. Use an antenna modeling program to calculate field strength at accessible locations.
D. All of the choices are correct.

When laying out your ham radio installation, remember it is the antenna that emits the most RF energy, and *antenna modeling on your computer can help you calculate field strengths* at any location where someone might get into the antenna pattern. Antenna modeling programs are available on the internet. **ANSWER C.**

E0A08 What does SAR measure?

A. Synthetic Aperture Ratio of the human body.
B. Signal Amplification Rating.
C. The rate at which RF energy is absorbed by the body.
D. The rate of RF energy reflected from stationary terrain.

SAR stands for *Specific Absorption Rate, the rate at which RF energy is absorbed by the human body*. **ANSWER C.**

E0A02 When evaluating RF exposure levels from your station at a neighbor's home, what must you do?

A. Make sure signals from your station are less than the controlled MPE limits.
B. Make sure signals from your station are less than the uncontrolled MPE limits.
C. You need only evaluate exposure levels on your own property.
D. Advise your neighbors of the results of your tests.

Your *neighbor's house is considered "uncontrolled"* when calculating maximum permissible exposure limits. Always make sure you do the calculations before putting up your new antenna system, and make absolutely sure you are always *less than the uncontrolled MPE limit*. **ANSWER B.**

E0A06 Why are there separate electric (E) and magnetic (H) field MPE limits?

A. The body reacts to electromagnetic radiation from both the E and H fields.
B. Ground reflections and scattering make the field impedance vary with location.
C. E field and H field radiation intensity peaks can occur at different locations.
D. All of these choices are correct.

Radio transmissions contain both an electric field as well as a magnetic field. These field radiations may peak at different distances from the antenna. In addition, ground reflections and scattering will vary the field intensity. And, the human body reacts differently to E and H fields. So take *all of these factors* into account when do your calculations for *both magnetic and electric fields*. **ANSWER D.**

E0A01 What, if any, are the differences between the radiation produced by radioactive materials and the electromagnetic energy radiated by an antenna?

A. There is no significant difference between the two types of radiation.
B. Only radiation produced by radioactivity can injure human beings.
C. Radioactive materials emit ionizing radiation, while RF signals have less energy and can only cause heating.
D. Radiation from an antenna will damage unexposed photographic film but ordinary radioactive materials do not cause this problem.

Radioactive radiation, like that used at hospitals for cancer treatment, breaks apart atoms and molecules in certain cells, and is usually administered in precise areas of the body. RF radiation is what comes out of microwave transmitters and your microwave oven, only heating molecules. Your transmitting antenna *signals have less energy* than radioactive materials, and *radio frequency emissions can only cause heating*. Radioactive materials emit ionizing radiation which is altogether different from radio signal output. **ANSWER C.**

E9C12 When constructing a Beverage antenna, which of the following factors should be included in the design to achieve good performance at the desired frequency?

A. Its overall length must not exceed 1/4 wavelength.
B. It must be mounted more than 1 wavelength above ground.
C. It should be configured as a four-sided loop.
D. It should be one or more wavelengths long .

The *Beverage antenna* is a tremendous signal-grabber and radiator, but for best performance, it really *needs to be L O N G*! The longer the better for a Beverage antenna. **ANSWER D.**

E9H09 Which of the following describes the construction of a receiving loop antenna?

A. A large circularly-polarized antenna.
B. A small coil of wire tightly wound around a toroidal ferrite core.
C. One or more turns of wire wound in the shape of a large open coil.
D. A vertical antenna coupled to a feed line through an inductive loop of wire.

A receiving loop antenna, for ham radio receive purposes, is usually many turns of wire wound in the shape of a large open coil or square loop. The more turns you add, the greater the gain of the receiving loop antenna. Most modern, high-frequency transceivers offer a separate antenna port for your homebrew "one or more turns of wire" *large open-coil antenna*. **ANSWER C.**

E9H10 How can the output voltage of a multi-turn receiving loop antenna be increased?

A. By reducing the permeability of the loop shield.

B. By increasing the number of wire turns in the loop and reducing the area of the loop structure.

C. By winding adjacent turns in opposing directions.

D. By increasing either the number of wire turns in the loop or the area of the loop structure or both.

To make a loop antenna more powerful, *either increase the number of wire turns* in the loop *or make the loop physically larger*. **ANSWER D.**

E9H11 What characteristic of a cardioid-pattern antenna is useful for direction finding?

A. A very sharp peak.

B. A very sharp single null.

C. Broad band response.

D. High-radiation angle.

A *cardioid pattern* is shaped like a heart. There is a broad reception pattern in one direction and a *deep null* in the other. When the button is pushed to "sense" the direction of the incoming signal on a direction-finding system that has a cardioid antenna pattern, the deep null in the pattern is normally 180 degrees away from the incoming signal. It's easier to use the null point to sense the signal direction of the incoming signal than it is the broad pattern – look for the minimum signal, not the maximum. **ANSWER B.**

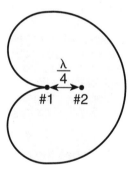

Two 1/4 wave vertical antennas fed 90-degrees out of phase will have a cardioid radiation pattern. The deep null will be 180-degrees from the transmitter, and is very helpful in direction-finding.

E9H12 What is an advantage of using a shielded loop antenna for direction finding?

A. It automatically cancels ignition noise pickup in mobile installations.

B. It is electro-statically balanced against ground, giving better nulls.

C. It eliminates tracking errors caused by strong out-of-band signals.

D. It allows stations to communicate without giving away their position.

Transmitter hunting down on 75 meters is an international sport. Most operators use a shielded loop antenna for T-hunting because the *shielding offers better directivity*. **ANSWER B.**

E9H05 What is the main drawback of a wire-loop antenna for direction finding?
  A. It has a bidirectional pattern.
  B. It is non-rotatable.
  C. It receives equally well in all directions.
  D. It is practical for use only on VHF bands.
Wire loop direction finding antennas will pick up maximum signal strength when the loop is broadside to the signal source. It is bi-directional – it will *pick up noise sources on either side of the loop.* **ANSWER A.** ☞ **Visit: www.lwca.org**

E9H07 Why is it advisable to use an RF attenuator on a receiver being used for direction finding?
  A. It narrows the bandwidth of the received signal to improve signal to noise ratio.
  B. It compensates for the effects of an isotropic antenna, thereby improving directivity.
  C. It reduces loss of received signals caused by antenna pattern nulls, thereby increasing sensitivity.
  D. It prevents receiver overload which could make it difficult to determine peaks or nulls.
When using a receiver for direction finding, a step adjustable or variable adjustable attenuator is necessary on the input when you get extremely close to the signal source. You *increase attenuation* in order *to keep your receiver from overloading* as you maintain maximum sensitivity for signal variations as the antenna direction is changed. **ANSWER D.** ☞ **Visit: www.homingin.com**

E9H08 What is the function of a sense antenna?
  A. It modifies the pattern of a DF antenna array to provide a null in one direction.
  B. It increases the sensitivity of a DF antenna array.
  C. It allows DF antennas to receive signals at different vertical angles.
  D. It provides diversity reception that cancels multipath signals.
On high frequency automatic radio direction finder systems, the double loop may employ an additional rigid whip to act as a *sense antenna*. Push the ADF sense button and the fixed double loop DF (direction finding) antenna *will null out* ambiguity as to which way the incoming signal is arriving from at that double loop. The sense antenna must be rigid, and never a flexing whip. **ANSWER A.**

E9H06 What is the triangulation method of direction finding?

A. The geometric angle of sky waves from the source are used to determine its position.
B. A fixed receiving station plots three headings from the signal source on a map.
C. Antenna headings from several different receiving locations are used to locate the signal source.
D. A fixed receiving station uses three different antennas to plot the location of the signal source.

Amateur radio operators participating in a joint "T-hunt" will swing their beams and record a magnetic bearing to where signal strength is the greatest. *Through triangulation, they can* then *determine the approximate location of the transmitting station*. All participating stations should agree to use either magnetic north or true north for their heading directions. **ANSWER C.**

Gordo and Chip, K7JA, fox-hunting with a 2-meter cubical quad antenna.

E0A07 How may dangerous levels of carbon monoxide from an emergency generator be detected?

A. By the odor.
B. Only with a carbon monoxide detector.
C. Any ordinary smoke detector can be used.
D. By the yellowish appearance of the gas.

While you might be able to smell generator exhaust, what you can't smell or see is a generator byproduct, carbon monoxide. Maybe you are working Field Day inside your motor home, and you have the generator exhaust well clear of the vehicle chassis. However, any small leak in the structure between your operating location and the generator could allow that generator to put CO into your position. *Only a carbon monoxide detector* will alert you that you are running out of good air! **ANSWER B.**

4

Taking the Extra Class Examination

Get set to pass the Element 4 examination and gain additional kilohertz of band privileges as an Extra Class operator! If you are upgrading from General to Extra Class, you will gain a whopping 500 kHz of *additional* frequencies on the 75/80-meter, 40-meter, 20-meter, and 15-meter worldwide service bands. If you are upgrading from a grandfathered Advanced Class license, you will gain an *additional* 250 kHz of operating area on the 75/80-meter, 40-meter, 20-meter, and 15-meter worldwide amateur service bands. Your Extra Class license gives you full privileges and full frequency coverage on every amateur service band assigned by the Federal Communications Commission.

In this chapter, we'll tell you how to prepare yourself for the exam, what to expect when you take the exam, and how you'll get your new, upgraded Extra Class license.

EXAMINATION ADMINISTRATION

Three Extra Class accredited Volunteer Examiners are required to administer your Element 4 exam. Many of you preparing for your Extra Class examination are already part of a Volunteer Examination team. If you are, it's perfectly acceptable to allow your fellow team members to administer a computer-generated, fresh Extra Class written theory examination.

Let's take a moment to review how well the Volunteer Examination program is working. The examination sessions are coordinated by national or regional Volunteer Examiner Coordinators (VECs) who accredit General Class, Advanced Class and Extra Class amateur operators as Volunteer Examiners (VEs). Your Extra Class examination **must** be taken at an official examination session in front of a team of three accredited Extra Class VEs.

The Volunteer Examiners are not compensated for their time and skills, but they are permitted to charge you an examination fee for certain reimbursable expenses incurred in preparing, administering, and processing the examination. The maximum fee is adjusted annually based on inflation by the VEC organizations. It is currently about $14.00.

HOW TO FIND AN EXAM SITE

Exam sessions are held regularly at sites throughout the country to serve their local communities. The exam site could be a public library, a fire house, someone's office, in a warehouse, and maybe even in someone's private home. Each examination team may regularly post their examination locations down at the local ham radio store. They also inform their VEC when and where they regularly hold test sessions. So the easiest way for you to find an exam session that is near you and at a convenient time is to call the VEC.

A complete list of VECs is located on page 269 in the Appendix of this book. The W5YI VEC and the ARRL VEC are the two largest examination groups in the country. They test in all 50 states and in foreign countries, too. Their 3-member, accredited examination teams are just about *everywhere*. So when you call the W5YI-VEC in Texas, or the ARRL-VEC in Connecticut, be assured they probably have an examination team only a few miles from where you are reading this book right now!

Want to find a test site fast?
Visit the W5YI-VEC website at www.w5yi.org, or call them at 800-669-9594.

Any of the VECs listed will provide you with the phone number of a local Volunteer Examiner team leader who can tell you the schedule of upcoming exam sessions near you. Select a session you wish to attend, and make a reservation so the VEs know to expect you. Don't be a "no-show" and don't be a "surprise-show." Make a reservation. And don't hesitate to tell them how much we all appreciate their efforts in supporting ham radio testing.

Ask them ahead of time what you will need to bring to the examination session. And ask them how much the current fee is for your exam session.

WHAT TO BRING TO THE SITE

Here is what you'll need to bring with you to the exam site for your upgrade from General or grandfathered Advanced to Extra Class:

1. The original and 1 copy of your present General or grandfathered Advanced Class license, or any CSCE issued within the last 365 days confirming examination credit. If you show up without your present license or copies of your present license, your examination or examination results could be put on a major hold.

2. A picture ID. Usually your Driver's License will suffice. Kids should bring their school ID, and very young kids should bring an adult to verify that they are who they are!

3. An examination fee of approximately $14.00 in cash, *exact change, please*. Personal checks will not be accepted, and plastic is out of the question. Ask the VE how much the fee will be when you call to make your exam reservation.

4. If you have a physical disability that may require additional assistance by your examination team, let them know ahead of time so they may prepare an examination that accommodates your special requirements. This could be special reading equipment, a volunteer examiner reader, access for a wheelchair, or any other special requirements that they may be able to set up for you. But they must know ahead of time what requirements need to be met, so please contact the volunteer examiner team leader in advance.

5. You will definitely want to **bring your calculator**. Calculators are indeed permitted and quite necessary for the Extra Class exam. Since some of the questions are from the old Advanced test, you can be sure you will have at least 8 or 10 formula problems to work on the calculator keys. The Volunteer Examination team leader may ask you to remove the battery from the calculator in order to verify memory dump. Don't be surprised if they ask you to completely clear your calculator in this fashion.

6. Bring plenty of sharp pencils and a fine-tip pen. Have backups. No red ink, please.

Try to arrive at the examination location about an hour before the exam begins. Introduce yourself to the Volunteer Examiners who will be conducting the test, and see if they need a hand in setting up the room. If it is a morning examination, Gordon West Radio School students usually bring donuts, hot coffee, tea, and hot chocolate for the Volunteer Examiners as a way of expressing their appreciation for the time and effort the examiners expend arranging and giving these tests. Start their day off with a smile and go for a dozen donuts! Believe me, they will appreciate the thought.

RUMOR MILL

While standing in line ready to enter the examination room, rumors usually begin to fly. Someone who is ill-informed may say they are still requiring the 20-wpm code test for Extra. Nonsense!

Someone may remark that all of the test questions have been changed, and this "new" 2016 test is different than the book. Nonsense! Again, the questions in this Extra Class book are valid from July 1, 2012, through June 30, 2016.

Someone comes out of the exam room indicating the Extra Class test had numerous questions not found in his book. Again, ridiculous! This book has every single test question, exactly as they will appear on the test. Only the A, B, C, and D order of the possible answers may be scrambled.

TAKING THE EXAMINATION

Get a good night's sleep before the exam day. Continue to study the Extra Class formulas right up to the moment you go into the examination room. Don't believe that old saying that too much study will cause you to forget the subject material. If you purchased my audio CD course, continue to play them up to the last minute.

Listen carefully for the instructions of your Volunteer Examiners. Most examination sessions will conduct all levels of testing, so stay tuned in on the announcements they give on who should come forward to pick up what type of test materials.

Make no marks on the examination sheet. If you don't know an answer, slightly shade in one of the answer sheet boxes, and make yourself a big note to go back and work out the correct answer. Using this technique will keep you from accidentally skipping a hard question, and misaligning your answer boxes with the easier questions. If you finish the exam at question #49, chances are you skipped a question and all of your work will be out of synch with your answer sheet. Double check that all 50 answer boxes indeed have a mark.

When you are taking the written examination, *take your time!* Some answers start out looking correct, but end up wrong at the end. Don't speed read the exam – getting careless after all this preparation could undo all the hard work you have put in on preparing for the Extra Class exam.

If you don't know an answer, eliminate the obviously wrong answers, and then take your best guess. NEVER LEAVE AN ANSWER BLANK! Leaving a specific answer blank is counted as a wrong answer. We also find that going back and second-guessing an answer will sometimes get you into the wrong answer category. Usually your first impression is the correct one. Another trick is to read your answer first and then see if it agrees with what the question asks. Start this technique at Answer 50, and go backwards.

AND THE RESULTS ARE...

When you are satisfied that you have passed the Extra Class exam, turn in all of your examination papers – including scratch paper and the original test paper. Thank your examiners for their courtesy, and follow their instructions while they prepare to score your paper. *Don't stand over the examiners* as they grade your test.

When you pass the test, the examiners may or may not indicate how many questions you missed. You missed only one? But which one? DON'T ASK – the examiners may tell you it is against the rules to divulge which question you got wrong, although we're not aware of any rule that states they cannot tell you what you got wrong. It's usually just a case of not enough time to spend with any one applicant pouring over their test performance. They usually will tell you your test score, and passing the exam is exactly what you want to hear!

Since we know you are going to be successful in your Extra Class written exam, we know that your examiners will issue you an official CSCE – Certificate of Successful Completion of Examination. This is an important document that is official proof that you have passed the exam. Check to see that it is signed by all three volunteer examiners. Did they get your name and call sign correct on the certificate? Write in your address if all they have is your name on the certificate, and be sure to sign your CSCE. Double-check that the date is correct. If the computer system should gobble up all of your test results, this is LEGAL PROOF that you indeed passed a particular examination element.

COMPLETING NCVEC FORM 605

Your examiners will assist you in filling out the NCVEC Form 605. They also will have copies of the most current version of this form at the examination session. (See sample form on page 272.) *Be aware that this form is **not** FCC form 605*, which is not suitable for use when taking an amateur radio examination or for changing an address, name, routine call sign change, or renewing your license through a VEC. Note that NCVEC Form 605 asks for your Social Security Number. Look at the copy of your current ham radio license and spot the numbers that show your Federal Registration Number (FRN). These numbers go in place of where they ask for your Social Security Number on the application. If this is your FIRST ham radio test, along with Technician and General Class, you will initially need to enter your Social Security number. If you have a current callsign, you most likely have the FRN, and this number and this number goes in place of your SSN

> **NOTE:** The renewal period on your new Extra Class license is NOT 10 years, but rather the SAME renewal date now on your General Class license. Make sure your address is up-to-date, too!

When you are filling-out NCVEC Form 605, it is important to compare the information on your present license with what you are writing on the Form 605. Any difference in your name or address requires that you check the appropriate CHANGE box. For example, if your current license reads "Jack" as your first name and you write down "John" on Form 605, your application will automatically be rejected when it is filed electronically. Same thing for your middle initial – does it appear on your original license? And make certain that your signature matches your name. Be sure to date the application.

Only fill out the top portion (Section 1) of NCVEC Form 605 when taking an examination. Section 2 (certifying section) is completed by the Volunteer Examiners after you complete the examination.

CHANGING CALL SIGNS

After passing the Extra Class exam, you have an opportunity to change your call sign systematically. Most examinees keep the one they have. If you check the box "CHANGE my station call sign systematically," the FCC computer will assign you the next Group "A" call sign from an alphabetized list.

The key word is *systematically*. You will not get to choose a specific call sign at an examination session. If you live in the continental United States, your assigned Group "A" systematic call sign will be in a 2-by-2 format with the first letter being

Only check the "CHANGE" my station call sign box if you want the FCC computer to give you a new call sign. If you want to choose a call sign, apply for a vanity call.

"A." A 2-by-2 format has two letters followed by a numeral and two more letters, such as AK6CS. Only the AA through AL block is assigned since AM through AZ 2-by-2 letters are not internationally allocated for use in the U.S. Once all 2-by-2 Group "A" call signs are assigned (the last will be AK1ZZ if you are in the first call district), the FCC then assigns call signs from the next lower (Group B) level to Extra Class amateurs. These will be 2-by-2 call signs beginning with K, N and W.

You must initial the request if you decide to systematically change your call sign. Once changed, it cannot easily be changed back to your previous call sign. The only way you can go back to a previously-held call sign is under the Vanity call sign program and this will cost you a small fee. So be thinking about it right now – do you want to stay with your present call sign, or do you want to change over to a new Extra Class 2-by-2 call sign?

VANITY CALL SIGNS

The ability to select a vacant station call sign of your choice began in 1996 and it has been immensely popular, especially with Extra Class amateurs who historically like to have short call signs. While only a little more than 10% of all radio amateurs ever make it to the top Extra Class level, they hold nearly half of all Vanity call signs issued!

Extra Class radio amateurs can choose an available station call sign containing any Group, A, B, C or D Amateur format. Most try to get a 1-by-2 or a 2-by-1 format, which are only available to an Extra Class licensee. A 1-by-2 Group "A" call sign begins with either K, N or W (but not "A") followed by a numeral and two letters. For example: W5YI is a Group "A" call sign.

An example of a 2-by-1 call sign is AA1A. The two letter prefix must be from the AA to AL, KA to KZ, NA to NZ or the WA to WZ prefix block followed by a numeral and a single letter. To make matters more confusing, certain prefixes (AH, AL, KH, KL, KP, NH, NL, NP, WH, WL and WP) are reserved for amateurs who have a mailing address outside of the 48 contiguous U.S. states.

You can find out all the rules (*and there are many!*) surrounding obtaining a Vanity call sign by visiting the W5YI Group website at www.w5yi.org and clicking on the "Vanity Call Signs" button near the top of the home page.

There is an additional regulatory fee associated with a Vanity call sign of about $12.00. You can complete FCC Form 605 (*not* NCVEC Form 605) and accompanying Form 159 (Remittance Advice) and file the application yourself either using paper documents or online. Additional information on Vanity call signs is available from the FCC's Amateur Radio website located at:

http://wireless.fcc.gov/services/amateur/licensing/vanity.html

---

An easier way to get a Vanity call sign is to let the W5YI Group handle everything for you. They charge a small, additional fee and do all the paperwork and filing for you. You can obtain complete details on this service by going to:
www.w5yi.org/vanity.htm

As for me, I'm staying with my original-issue WB6NOA call sign. If I changed it, I would be breaking a 50+ year tradition.

GOING FURTHER - GROL + RADAR

The Extra Class license is the highest level license you can achieve in the Amateur Radio service. Did you know you can also take the commercial general radio operator (GROL) license exams, too? The commercial license allows you to work on marine and aeronautical two-way radio equipment, and the commercial GROL license often is required by two-way radio employers before being hired.

And guess what? The GROL technical element 3 uses many identical test questions as those you have just studied for your amateur Extra Class license. In fact, except for just a couple of numbers, the commercial radiotelephone examination may use identical math problems that you just studied.

So, why not upgrade to a professional-level FCC license? You can order the *GROL + RADAR* book by calling the W5YI Group at 800-669-9594.

BECOMING A VOLUNTEER EXAMINER

When you successfully complete your Extra Class exam, ask your 3-member accredited Volunteer Examination team for a volunteer examiner sign-up application. It only takes a couple of weeks to process, and you'll be joining a cadre of thousands of other Extra Class hams who give something back to ham radio when it comes to testing – just a few hours a month, and some paper and computer processing. It is a rewarding "extra credit" as an Extra Class operator, and you and two other accredited examiners could even give tests abroad all over the world.

Accredited Volunteer Examiners also may help develop new questions when the amateur radio question pools undergo periodic updating. Your comments and suggestions will be filed with the prestigious NCVEC Question Pool Committee, and they will consider your valued contributions when each Element 2, Element 3, and Element 4 question pool is updated. Each pool gets updated once every 4 years with a public notice for radio experts to submit new or improved examination questions along with 4 suggested right and wrong answers. Join in – as an Extra Class, it's fun to see that you have contributed to the new question pool.

Here's how to become a volunteer examiner:
Write the W5YI-VEC at: P.O. Box 500065, Arlington, TX 76006-0065,
or call them at 817-860-3800 during regular business hours to obtain a VE application.
You also may apply online at: www.w5yi.org.
Click on the "Become a Volunteer Examiner" button.

SUMMARY

You made it through Extra Class! Congratulations! Allow me to send you a free certificate of achievement. Send a large envelope, self-address, with 12 first class stamps on the inside to me at Gordon West Radio School, 2414 College Drive, Costa Mesa, California 92626. It will take about 15 days to get the certificate back to you, and a hearty congratulations on making it to the top.

I hope to work you soon on the worldwide airwaves, or see you at an upcoming hamfest. Think about getting that commercial radio license, too, based on all of the study you have just completed for amateur Extra Class.

Welcome to the top!

73
Gordon West,
WB6NOA

**I know congratulations will be in order very soon
when you pass your Extra Class exam with flying colors!**

APPENDIX

U.S. VOLUNTEER EXAMINER COORDINATORS IN THE AMATEUR SERVICE

Anchorage Amateur Radio Club
P.O. Box 670616
Chugiak, AK 99567-0616
907/338-0662
e-mail: jwiley@alaska.net

ARRL/VEC
225 Main Street
Newington, CT 06111-1494
860/594-0300
860/594-0339 (fax)
e-mail: vec@arrl.org
Internet: www.arrl.org

Central America VEC, Inc.
Larry Frost KR4GU
2751 Christian LN NE
Huntsville, AL 35811-1864
256/288-0392
256/653-5007
e-mail: cavec@bellsouth.net

Golden Empire Amateur Radio Society
P.O. Box 508
Chico, CA 95927-0508
530/345-3515
e-mail: wa6zrt@sbcglobal.net

Greater Los Angeles
Amateur Radio Group
P.O. Box 500133
Palmdale, CA 93591
818/222-7013
818/892-9855 (fax)
vec@glaarg.org

Jefferson Amateur Radio Club
Keith Barnes W5KB
P.O. Box 73665
Metairie, LA 70033-3665
504/831-1613
e-mail: w5kb@w5gad.org
Internet: www.w5gad.org

Laurel Amateur Radio Club, Inc.
4708 Montgomery PL
Beltsville, MD 20705-2921
301/937-0394 (6-9 PM)
e-mail: aa3of@arrl.net
Internet: www.larcmdorg.doore.net/vec

The Milwaukee Radio Amateurs Club, Inc.
P.O. Box 070695
Milwaukee, WI 53207-0695
262/797-6722
e-mail: tom@supremecom.biz

MO-KAN/VEC
228 Tennessee Road
Richmond, KS 66080-9174
785/867-2011
e-mail: wo0e@lcwb.coop

SANDARC-VEC
P.O. Box 2446
La Mesa, CA 91943-2446
619/697-1475
e-mail: n6nyx@arrl.net

Sunnyvale VEC Amateur Radio Club, Inc.
P.O. Box 60307
Sunnyvale, CA 94088-0307
408/255-9000 (exam info 24 hours)
e-mail: vec@amateur-radio.org
Internet: www.amateur-radio.org

W4VEC
Rae Everhart K4SWN
P.O. Box 482
China Grove, NC 28023-0482
e-mail: raef@lexcominc.net
Internet: www.w4vec.com

Western Carolina WCARS
7 Skylyn Ct.
Asheville, NC 28806-3922
e-mail: bstewart@windstream.net
Internet: www.wcarsvec.org

W5YI-VEC
P.O. Box 500065
Arlington, TX 76006-0065
817/860-3800
800/669-9495
e-mail: w5yi-vec@w5yi.org
Internet: www.w5yi.org

2012-16 ELEMENT 4 Q&A CROSS REFERENCE

The following cross reference presents all 703 question numbers in numerical order included in the 2012-16 Element 4 Extra Class question pool, followed by the page number on which the question begins in this book. This will allow you to located specific questions by question number. Note: 1 question was deleted from the pool by the Question Pool Committee, resulting in an active pool of 702 questions. The deleted question does not appear in this book, but its number is listed here followed by the word deleted.

INSTRUCTIONS FOR COMPLETING APPLICATION FORM NCVEC FORM 605

AMATEUR RADIO LICENSE LEVELS & REQUIREMENTS

As of April 15, 2000, you may be examined on only three classes of operator licenses, each authorizing varying levels of privileges. Currently there are three license classes; Technician, General and Extra. The Federal Communications Commission (FCC) grants these licenses.

Technician Class License
EXAM: 35-question Technician Written Exam (Element 2) PRIVILEGES: All VHF/UHF amateur bands (frequencies above 30 MHz) and certain HF frequencies using the 80, 40, and 15 meter bands using CW, and on the 10 meter band using CW, voice, and digital modes.

General Class License (upgrade from Technician)
EXAM: 35-question General Written Exam (Element 3) PRIVILEGES: All VHF/UHF amateur bands and most HF privileges (10 through 160 meters).
In addition to the Technician privileges, General Class operators are authorized to operate on any frequency in the 160, 30, 17, 12, and 10meter bands. They may also use significant segments of the 80, 40, 20, and 15meter bands.

Extra Class License (upgrade from General)
EXAM: 50-question Extra Written Exam (Element 4) PRIVILEGES: All amateur privileges.

Should you have any questions, contact one of the 14 volunteer examiner coordinator (VEC) organizations on www.ncvec.org. To contact the FCC, call 888-225-5322 (weekdays), or write to FCC, 1270 Fairfield Road, Gettysburg PA 17325-7245. Also see the FCC website http://wireless.fcc.gov/services/index.htm?job=service_home&id=amateur

RENEWING OR MODIFYING YOUR AMATEUR RADIO STATION LICENSE

NCVEC FORM 605
The NCVEC Form 605 may also be used to renew or modify your Amateur Radio Operator/Primary Station license. License renewals (with associated fee for vanity renewals) may only be completed during the final 90 days prior to license expiration, or up to two years after expiration. Changes to your mailing address, name and requests for a sequential change of your station call sign appropriate for your license class may be requested at any time. This form may not be used to apply for a new specific "vanity" station call sign. Do not send or provide this form to the FCC – this form is for VE / VEC use only.

THE FCC APPLICATION FORM 605
The FCC version of the form 605 may not be used for requests submitted to a VE team or a VEC since it does not request information needed by the administering VE's. The FCC Form 605 may however, be used to routinely renew or modify your license without charge. New vanity license requests and vanity license renewals require a FCC regulatory fee. FCC form 605 should be sent to the FCC, 1270 Fairfield Rd, Gettysburg PA 17325-7245. FCC forms can be obtained via the FCC web at www.fcc.gov/formpage.html or by fax at 202-418-0177.

ARE WRITTEN TESTS AN FCC-LICENSE REQUIREMENT? ARE THERE EXEMPTIONS?

The Amateur Operator/Primary Station License class for which an examinee qualifies for is determined by the exams taken at a VE test session. The exams cover regulations, operating practices, and electronics theory. There is no exemption from the written exam requirements for a handicap or disability. There are exam accommodations that can be afforded examinees. Most new amateur operators start at the Technician class and then advance one class at a time. FCC rules provide for examination element credit in some cases, licenses that were previously held so that examinations required for that license need not be repeated. The written examinations are constructed from question pools that are in public domain, see: www.ncvec.org. Helpful study guides and training courses are also widely available. To locate examination opportunities in your area, contact your local club, VE group, one of the 14 VECs or see the online listings on www.ncvec.org.

IS MORSE CODE AN FCC-LICENSE REQUIREMENT?

Beginning February 23, 2007, the FCC no longer requires passing a Morse code examination for any Amateur Radio license class.

SOCIAL SECURITY NUMBER

Under the Debt Collection Act of 1996 your Taxpayer ID Number (TIN), which is your Social Security Number, is required on this application – or your FCC-assigned Federal Registration Number (FRN). An FRN is assigned by the FCC registration system as soon as your SSN is registered.

RENEWING OR MODIFYING YOUR AMATEUR LICENSE

You can submit your renewal or license modifications to FCC via the Internet at: https://www.fcc.gov/uls. If you are already registered in ULS and have obtained a FRN (Federal Registration Number), you can choose the "ONLINE FILING / LOG IN" link to perform your on-line transaction with the FCC. If you do not have a FRN, you must first register in the ULS by following the "New Users / Register" link and complete your registration information. You can then choose the "Online Filing / LOG IN" link. Direct any on-line filing or password questions to FCC Tech Support weekdays at 202-414-1250.

CLUB STATION CALL SIGN ADMINISTRATORS (CSCSA)

The NCVEC Form 605 is also used for the processing of applications for Amateur Service club and military recreation station call signs and for the modification of RACES stations. This form may not be used to apply for a new specific "vanity" station call sign. The Club Station Call Sign Administrators (CSCSA) are: ARRL/VEC 225 Main St., Newington, CT 06111, W4VEC POB 41 Lexington, NC 27293-0041 and the W5YI-VEC POB 200065, Arlington, TX 76006. Please return this form to one of these three CSCSAs.

NCVEC FORM 605
February 2010 - Page 2

NCVEC QUICK-FORM 605 APPLICATION FOR AMATEUR OPERATOR/PRIMARY STATION LICENSE

SECTION 1 - TO BE COMPLETED BY APPLICANT

PRINT LAST NAME	SUFFIX (Jr, Sr.)	FIRST NAME	INITIAL	STATION CALL SIGN (IF ANY)

MAILING ADDRESS (Number and Street or P.O. Box)		SOCIAL SECURITY NUMBER (SSN) or (FRN) FCC FEDERAL REGISTRATION NUMBER

CITY	STATE CODE	ZIP CODE (5 or 9 Numbers)	E-MAIL ADDRESS (OPTIONAL)

DAYTIME TELEPHONE NUMBER (Include Area Code) OPTIONAL	FAX NUMBER (Include Area Code) OPTIONAL	ENTITY NAME (IF CLUB, MILITARY RECREATION, RACES)

Type of Applicant:	Individual	Amateur Club	Military Recreation	RACES (Modify Only)	CLUB, MILITARY RECREATION, or RACES CALL SIGN
					SIGNATURE OF RESPONSIBLE CLUB OFFICIAL (not license)

I HEREBY APPLY FOR (Make an X in the appropriate box(es))

☐ EXAMINATION for a new license grant
☐ EXAMINATION for upgrade of my license class
☐ CHANGE my name on my license to my new name
Former Name: _____ (Last name) (First name) (MI)

☐ CHANGE my mailing address to above address
☐ CHANGE my station call sign systematically
Applicant's Initials: _____
☐ RENEWAL of my license grant.

Do you have another license application on file with the FCC which has not been acted upon?	PURPOSE OF OTHER APPLICATION	PENDING FILE NUMBER (FOR VEC USE ONLY)

I certify that:
· I waive any claim to the use of any particular frequency regardless of prior use by license or otherwise;
· All statements and attachments are true, complete and correct to the best of my knowledge and belief and are made in good faith;
· I am not subject to a denial of Federal benefits pursuant to Section 5301 of the Anti-Drug Abuse Act of 1988, 21 U.S.C. § 862;
· The construction of my station will NOT be an action which is likely to have a significant environmental effect. (See 47 CFR Sections 1.1301-1.1319 and Section 97.13(a));
· I have read and WILL COMPLY with Section 97.13(c) of the Commission's Rules regarding RADIOFREQUENCY (RF) RADIATION SAFETY and the amateur service section of OST/OET Bulletin Number 65.

Signature of applicant (Do not print, type, or stamp. Must match applicant's name above.) (Clubs: 2 different individuals must sign)

X _____ Date Signed: _____

SECTION 2 - TO BE COMPLETED BY ALL ADMINISTERING VEs

DATE OF EXAMINATION SESSION	EXAMINATION SESSION LOCATION

Applicant is qualified for operator license class:

		EXAMINATION SESSION	VEC ORGANIZATION
☐ NO NEW LICENSE OR UPGRADE WAS EARNED			
☐ TECHNICIAN	Element 2		VEC RECEIPT DATE
☐ GENERAL	Elements 2 and 3		
☐ AMATEUR EXTRA	Elements 2, 3 and 4		

I CERTIFY THAT I HAVE COMPLIED WITH THE ADMINISTERING VE REQUIRMENTS IN PART 97 OF THE COMMISSION'S RULES AND WITH THE INSTRUCTIONS PROVIDED BY THE COORDINATING VEC AND THE FCC.

1st VEs NAME (Print First, MI, Last, Suffix)	VEs STATION CALL SIGN	VEs SIGNATURE (Must match name)	DATE SIGNED
2nd VEs NAME (Print First, MI, Last, Suffix)	VEs STATION CALL SIGN	VEs SIGNATURE (Must match name)	DATE SIGNED
3rd VEs NAME (Print First, MI, Last, Suffix)	VEs STATION CALL SIGN	VEs SIGNATURE (Must match name)	DATE SIGNED

DO NOT SEND THIS FORM TO FCC – THIS IS NOT AN FCC FORM.
IF THIS FORM IS SENT TO FCC, FCC WILL RETURN IT TO YOU WITHOUT ACTION.

NCVEC FORM 605 - February 2010
FOR VE/VEC USE ONLY - Page 1

FCC Form 605 is used to process routine license renewals and license changes that are submitted directly to the FCC. The FCC will not accept NCVEC Form 605.

NCVEC Form 605 is used to process test results, license upgrades, call sign changes, and changes of address submitted through a VEC. VECs will not accept FCC Form 605.

SCHEMATIC SYMBOLS

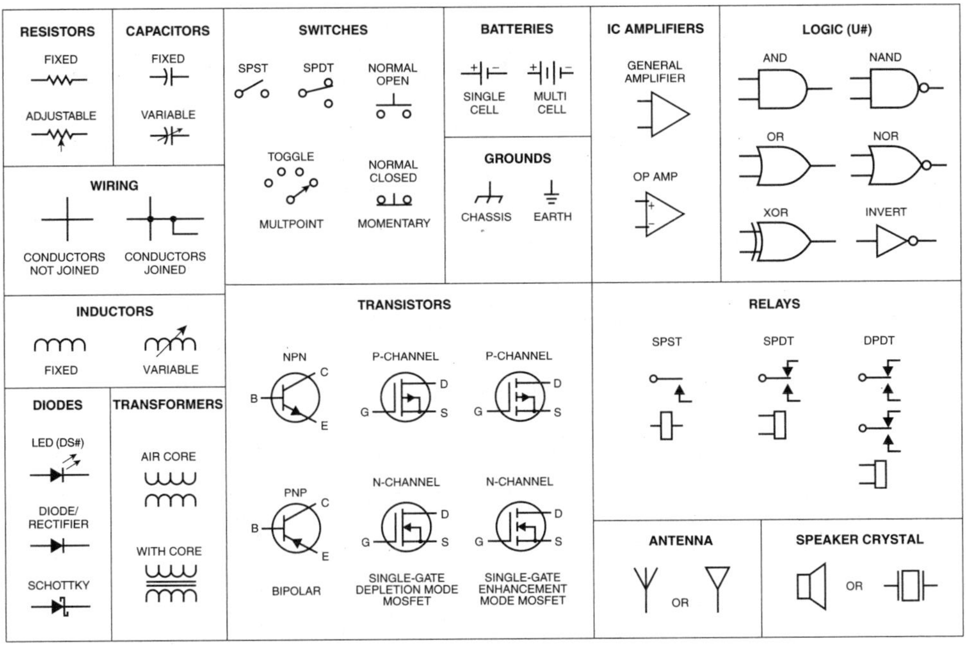

RESISTORS	CAPACITORS
FIXED	FIXED
ADJUSTABLE	VARIABLE

WIRING

CONDUCTORS NOT JOINED CONDUCTORS JOINED

INDUCTORS

FIXED VARIABLE

DIODES | **TRANSFORMERS**

LED (DS#)

AIR CORE

DIODE/ RECTIFIER

WITH CORE

SCHOTTKY

SWITCHES

SPST SPDT NORMAL OPEN

TOGGLE NORMAL CLOSED

MULTPOINT MOMENTARY

TRANSISTORS

NPN P-CHANNEL P-CHANNEL

C B E

G D S

PNP N-CHANNEL N-CHANNEL

BIPOLAR SINGLE-GATE DEPLETION MODE MOSFET SINGLE-GATE ENHANCEMENT MODE MOSFET

BATTERIES

SINGLE CELL MULTI CELL

GROUNDS

CHASSIS EARTH

IC AMPLIFIERS

GENERAL AMPLIFIER

OP AMP

LOGIC (U#)

AND NAND

OR NOR

XOR INVERT

RELAYS

SPST SPDT DPDT

ANTENNA

OR

SPEAKER CRYSTAL

OR

SCIENTIFIC NOTATION

Prefix	Symbol		Multiplication Factor
exa	E	10^{18} =	1,000,000,000,000,000,000
peta	P	10^{15} =	1,000,000,000,000,000
tera	T	10^{12} =	1,000,000,000,000
giga	G	10^{9} =	1,000,000,000
mega	M	10^{6} =	1,000,000
kilo	k	10^{3} =	1,000
hecto	h	10^{2} =	100
deca	da	10^{1} =	10
(unit)		10^{0} =	1

Prefix	Symbol		Multiplication Factor
deci	d	10^{-1} =	0.1
centi	c	10^{-2} =	0.01
milli	m	10^{-3} =	0.001
micro	μ	10^{-6} =	0.000001
nano	n	10^{-9} =	0.000000001
pico	p	10^{-12} =	0.000000000001
femto	f	10^{-15} =	0.000000000000001
atto	a	10^{-18} =	0.000000000000000001

SUMMARY OF QUESTION POOL FORMULAS

Page(s)	Formula	Where:
224-229 235-236 254	**Decibels** $dB = 10 \log_{10} \dfrac{P_1}{P_2}$	P_1 = Power (usually output) in **watts** P_2 = Power (usually input) in **watts**
215-216	**Counter Readout Error** Readout Error = f × a	f = Frequency in **MHz** being measured a = Counter accuracy in **parts per million**
124	**Intermodulation Product** $f_{IMD(2)} = 2f_1 - f_2$ $f_{IMD(4)} = 2f_2 - f_1$	f_{IMD} = Intermodulation product at particular f in **Hz** f_1 = Frequency of transmitted signal in **Hz** f_2 = Possible frequency of interfering signal in **Hz**
149 168-169	**Resonance** $X_L = X_C$ $2\pi f_r L = \dfrac{1}{2\pi f_r C}$ $f_r = \dfrac{1}{2\sqrt{LC}}$	X_L = Inductive reactance in **ohms** X_C = Capacitive reactance in **ohms** f_r = Resonant frequency in **hertz** L = Inductance in **henrys** C = Capacitance in **farads** π = 3.14
152	**Phase Between V and I** ELI = Voltage (E) leads current (I) in an inductive (L) circuit ICE = Current (I) leads voltage (E) in a capacitive (C) circuit	E = Voltage in circuit I = Current in circuit L = Inductance in circuit C = Capacitance in circuit
167-168	**Half-Power Bandwidth** $BW_{-3dB} = \dfrac{f_r}{Q}$	BW_{-3dB} = Half-Power bandwidth in **kHz** f_r = Resonant frequency in **kHz** Q = Circuit **quality factor**
169-173	**Time Constants** $\tau = RC$ $\tau = \dfrac{L}{R}$	τ = Time Constant in **seconds** R = Resistance in **ohms** C = Capacitance in **farads** L = Inductance in **henries**

Time Constant Curve

Time Constant τ	% of Change in τ	RC		RL	
		% of Final Q or V on C When Charging	% of Initial Q or V on C When Discharging	% of Final I When Increasing	% of Initial I When Decreasing
1	63.2	63.2	36.8	63.2	36.8
2	23.3	86.5	13.5	86.5	13.5
3	8.5	95.0	5.0	95.0	5.0
4	3.2	98.2	1.8	98.2	1.8
5	1.1	99.3	0.7	99.3	0.7

SUMMARY OF QUESTION POOL FORMULAS (continued)

Page(s)	Formula	Where:

170	**Capacitors** **Resistors**	

Series: $C_T = \dfrac{C_1 \times C_2}{C_1 + C_2}$ $R_T = R_1 + R_2$

C = Capacitance in **farads**

R = Resistance in **ohms**

Parallel: $C_T = C_1 + C_2$ $R_T = \dfrac{R_1 \times R_2}{R_1 + R_2}$

151-152	**Inductive Reactance**	X_L = Inductive reactance in **ohms**

$$X_L = 2\pi fL$$

L = Inductance in **henries**

f = Frequency in **hertz**

π = 3.14

151-152	**Capacitive Reactance**	X_C = Capacitive reactance in **ohms**

C = Capacitance in **farads**

$$X_C = \dfrac{1}{2\pi fC}$$

f = Frequency in **hertz**

π = 3.14

$$X_C = \dfrac{10^6}{2\pi fC}$$

f = Frequency in **MHz**

C = Capacitance in **picofarads**

Z = Impedance in **ohms**

Impedance in Series

π = 3.14

$$Z_T = Z_1 + Z_2$$

Impedance in Parallel ⟶ Special Case:

When $Z_1 = R$ and $Z_2 = \pm jX$

$$+Z_T = \dfrac{Z_1 \times Z_2}{Z_1 + Z_2}$$

$$Z_T = \dfrac{RX\ \underline{/\Theta = \pm 90^\circ}}{\sqrt{R^2 + X^2}\ \underline{/\text{arctan} \pm X/R}}$$

155-156	**Rectangular Coordinates**	Z = Impedance in **ohms**

$$Z = R \pm jX$$

R = Resistance in **ohms**

$+jX$ = Inductive reactance X_L in **ohms**

Polar Coordinates

$-jX$ = Capacitive reactance X_C in **ohms**

$$Z = Z\ \underline{/\pm\Theta}$$

Θ = Phase angle in **degrees**

arctan = angle whose tangent is

Conversion:

$Z = R \pm jX$ to $Z\ \underline{/\pm\Theta}$ $Z = \sqrt{R^2 + X^2}\ Z\ \underline{/\text{arctan}\ \frac{X}{R}}$

$Z\ \underline{/\pm\Theta}$ to $Z = R \pm jX$ $R = Z \cos\Theta$

$\pm jX = Z \sin\Theta$

152-155	**Phase Angle Between V and I**	ϕ = Phase angle in **degrees**

$X = (X_L - X_C)$ = Total reactance in **ohms**

$$\text{Tan } \phi = \dfrac{X}{R}$$

R = Resistance in **ohms**

SUMMARY OF QUESTION POOL FORMULAS (continued)

Page(s)	Formula	Where:

159-160

Do Addition and Subtraction in Rectangular Coordinates

$$Z_1 = R_1 + jX_1 \qquad Z_2 = R_2 - jX_2$$

Addition

$$Z_T = (R_1 + jX_1) + (R_2 - jX_2)$$
$$Z_T = (R_1 + R_2) + j(X_1 - X_2)$$

Subtraction

$$Z_T = (R_1 + jX_1) + (R_2 - jX_2)$$
$$Z_T = (R_1 - R_2) + j(X_1 + X_2)$$

Do Multiplication and Division in Polar Coordinates

$$Z_1 = Z_1 \,\underline{/\Theta_1} \qquad Z_2 = Z_2 \,\underline{/\Theta_2}$$

Multiply

$$Z_1 Z_2 = Z_1 \times Z_2 \,\underline{/\Theta_1 + \Theta_2}$$

Divide

$$\frac{Z_1}{Z_2} = \frac{Z_1 \,\underline{/\Theta_1}}{Z_2 \,\underline{/\Theta_1}} = \frac{Z_1}{Z_2} = \underline{/\Theta_1 - \Theta_2}$$

166

Circuit Quality Factor

$$Q = \frac{R}{X_L}$$

$$Q = \frac{R}{2\pi f_r L}$$

Q = Circuit **quality factor**
X_L = $2\pi f_r L$
X_L = Inductive reactance in **ohms**
f_r = Resonant frequency in **Hz**
L = Inductance in **henrys**
R = Resistance in **ohms**
2π = 6.28

176-178

True Power

$$P_T = P_A \times PF$$

$$P_A = E \times I$$

$$PF = \cos \phi$$

P_T = True power in **watts**
P_A = Apparent power in **watts**
ϕ = Phase angle in **degrees**
PF = Power factor
E = Applied voltage in **volts**
I = Circuit current in **amps**

296-208

Inverting IC Op-Amp Gain

$$G = -\frac{R_f}{R_1}$$

G = Gain in **non-dimentional units**
R_f = Feedback resistance in **ohms**
R_1 = Input resistance in **ohms**

179-181

Turns for L Using Ferrite Toroidal Cores

$$N = 1000 \sqrt{\frac{L}{A_L}}$$

N = Number of turns needed
A_L = Inductance index in **mH per 1000 turns**
L = Inductance in **millihenrys**

179-181

Turns for L Using Powdered-Iron Toroidal Cores

$$N = 100 \sqrt{\frac{L}{A_L}}$$

N = Number of turns needed
A_L = Inductance index in **µH per 100 turns**
L = Inductance in **microhenrys**

SUMMARY OF QUESTION POOL FORMULAS (continued)

Page(s)	Formula	Where:
94-95	**Peak, Peak-to-Peak and RMS Voltages** $$**V_P = \frac{V_{PP}}{2} \quad *V_{RMS} = 0.707\,V_P$$ $$**V_{PP} = 2V_P \quad *V_P = 1.414\,V_{RMS}$$	V_P = Peak voltage in **volts** V_{PP} = Peak-to-peak voltage in **volts** V_{RMS} = Root-mean-square voltage in **volts** * Equations for sinewaves ** Equations for symmetrical waves

92	**Peak Envelope Power** $PEP = P_{DC} \times$ Efficiency	PEP = Peak envelope power in **watts** P_{DC} = Input DC power in **watts**

Amplifier Class	Efficiency
C	80%
B	60%
AB	50%
A	<50%

91	**Modulation Index** $$\text{Modulation Index} = \frac{\text{Deviation of FM Signal (in Hz)}}{\text{Modulating Frequency (in Hz)}}$$

91-92	**Deviation Ratio** $$\text{Deviation Ratio} = \frac{\text{Maximum Carrier Frequency Deviation (in kHz)}}{\text{Maximum Modulation Frequency (in kHz)}}$$

87	**Bandwidth for CW** BW_{CW} = baud rate $\times$ wpm $\times$ fading factor	BW_{CW} = Necessary bandwidth in **hertz** wpm = Morse code signal rate in **words per minute** Fading factor = Constant of 5 for **CW**

81	**Bandwidth for Digital** BW = baud rate $+ (1.2 \times f_s)$	BW = Necessary bandwidth in **hertz** f_s = Frequency shift in **hertz** Baud rate = Digital signal rate in **bauds**

229	**Antenna Beamwidth** $$\text{Beamwidth} = \frac{203}{(\sqrt{10})^X}$$	Beamwidth = Antenna beamwidth in **degrees** A_G = Antenna gain in **dB** $$X = \frac{A_G}{10}$$

SUMMARY OF QUESTION POOL FORMULAS (continued)

Page(s)	Formula	Where:

243-248

Physical Length vs. Electrical Length

$$L = \frac{984\lambda V}{f}$$

L = Physical length in **feet**
λ = Electrical length in **wavelengths**
V = **Velocity factor** of feedline
f = Frequency in **MHz**

Feedline	Velocity Factor
Coax	0.66
Parallel	0.95
Twin-Lead	0.80

156

Complex Numbers (Real and Imaginary) and Operator j

Recall from algebra that $+2 \times +2 = +4$ and $-2 \times -2 = +4$ ∴ $\sqrt{4}$ has two roots $+2$ or -2. In ac circuit analysis there is a need to take the square root of a negative number. For example, $\sqrt{-4}$ has a root $\sqrt{-1} \times \sqrt{4} = j2$ and $-j2$. The root of a negative number is called an *imaginary number* and this is indicated by writing the operator j in front of the root. j is equal to $\sqrt{-1}$ and is the basic imaginary quantity. Since $j = \sqrt{-1}$, it is interesting to note the following because they will be encountered in ac circuit analysis:

$$j^2 = j \times j = \sqrt{-1} \times \sqrt{-1} = -1$$

$$j^3 = j \times j \times j = j^2 \times j = -1 \times j = -j$$

$$j^4 = j \times j \times j \times j = j^2 \times j^2 = -1 \times -1 = +1$$

$$\frac{1}{j} = \frac{1}{j} \times \frac{j}{j} = \frac{j}{j^2} = \frac{j}{-1} = -j$$

Summary

$$j = \sqrt{-1}$$
$$j^2 = -1$$
$$j^3 = -j$$
$$j^4 = +1$$
$$\frac{1}{j} = -j$$

j Operator as Vector Rotator

Note that an imaginary quantity can be considered as a vector being rotated by the operator j.

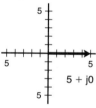

Vector starts at 0°.

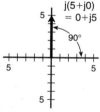

The j operator, when it multiplies a vector, rotates the vector by 90° counter clockwise.

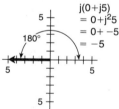

A second 90° rotation puts the vector at 180°.

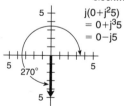

A third 90° rotation puts the vector at 270°.

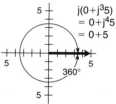

A fourth 90° rotation puts the vector back at 0°.

AUTHORIZED FREQUENCY BANDS – AMATEUR SERVICE (for U.S. Amateur Stations operating from ITU-Region 2–North and South America)

METERS	Current License Class[1]				
	Grandfathered[2]		General	Advanced	Extra Class
	Novice	Technician			
160			1800-2000 kHz/All	1800-2000 kHz/All	1800-2000 kHz/All
80 / 75	3525-3600 kHz/CW	3525-3600 kHz/CW	3525-3600 kHz/CW 3800-4000 kHz/Ph	3525-3600 kHz/CW 3700-4000 kHz/Ph	3500-4000 kHz/CW 3600-4000 kHz/Ph
60			5 specific channels[4]	5 specific channels[4]	5 specific channels[4]
40	7025-7125 kHz/CW	7025-7125 kHz/CW	7025-7125 kHz/CW 7175-7300 kHz/Ph	7025-7125 kHz/CW 7125-7300 kHz/Ph	7000-7300 kHz/CW 7125-7300 kHz/Ph
30			10.1-10.15 MHz/CW	10.1-10.15 MHz/CW	10.1-10.15 MHz/CW
20			14.025-14.15 MHz/CW 14.225-14.35 MHz/Ph	14.025-14.15 MHz/CW 14.175-14.35 MHz/Ph	14.0-14.35 MHz/CW 14.15-14.35 MHz/Ph
17			18.068-18.11 MHz/CW 18.11-18.168 MHz/Ph	18.068-18.11 MHz/CW 18.11-18.168 MHz/Ph	18.068-18.11 MHz/CW 18.11-18.168 MHz/Ph
15	21.025-21.2 MHz/CW	21.025-21.2 MHz/CW	21.025-21.2 MHz/CW 21.275-21.45 MHz/Ph	21.025-21.2 MHz/CW 21.225-21.45 MHz/Ph	21.0-21.45 MHz/CW 21.2-21.45 MHz/Ph
12			24.89-24.99 MHz/CW 24.93-24.99 MHz/Ph	24.89-24.99 MHz/CW 24.93-24.99 MHz/Ph	24.89-24.99 MHz/CW 24.93-24.99 MHz/Ph
10	28.0-28.5 MHz/CW 28.3-28.5 MHz/Ph	28.0-28.5 MHz/CW 28.3-28.5 MHz/Ph	28.0-28.3 MHz/CW 28.3-29.7 MHz/Ph	28.0-28.3 MHz/CW 28.3-29.7 MHz/Ph	28.0-29.7 MHz/CW 28.3-29.7 MHz/Ph
6		50-54 MHz/CW 50.1-54 MHz/Ph	50-54 MHz/CW 50.1-54 MHz/Ph	50-54 MHz/CW 50.1-54 MHz/Ph	50-54 MHz/CW 50.1-54 MHz/Ph
2		144-148 MHz/CW 144.1-148 MHz/All	144-148 MHz/CW 144.1-148 MHz/All	144-148 MHz/CW 144.1-148 MHz/All	144-148 MHz/CW 144.1-148 MHz/All
1.25	222-225 MHz/All	222-225 MHz/All[3]	222-225 MHz/All	222-225 MHz/All	222-225 MHz/All
0.70		420-450 MHz/All	420-450 MHz/All	420-450 MHz/All	420-450 MHz/All
0.33		902-928 MHz/All	902-928 MHz/All	902-928 MHz/All	902-928 MHz/All
0.23	1270-1295 MHz/All	1240-1300 MHz/All	1240-1300 MHz/All	1240-1300 MHz/All	1240-1300 MHz/All

1 Effective 4-15-00   2 Prior to 4-15-00   3 Effective 2/1/94 219-220 MHz is authorized for point-to-point fixed digital message forwarding systems.
4 60 meter operation restricted to 5 channels with center frequencies of 5332.0, 5348.0, 5358.5, 5373.0 and 5405.0 kHz, 100 watts PEP. See FCC 97.303.

Note: Morse code (CW, A1A) may be used on any frequency allocated to the amateur service. Telephony emission (abbreviated Ph above) authorized on certain bands as indicated. Higher class licensees may use slow-scan television and facsimile emissions on the Phone bands; radio teletype/digital on the CW bands. All amateur modes and emissions are authorized above 144.1 MHz. In actual practice, the modes/emissions used are somewhat more complicated than shown above due to the existence of various band plans and "gentlemen's agreements" concerning where certain operations should take place.

ELEMENT 4 (EXTRA CLASS) QUESTION POOL SYLLABUS

The syllabus used by the QPC for the development of the question pool is included here as an aid in studying the subelements and topic groups. Review the syllabus before you start your study to gain an understanding of how the question pool is used to develop the Element 4 written examination. Remember, one question will be taken from each topic group within each subelement to create your exam. The QPC topic groups are not the same as those used by Gordon West in this book.

E1 – COMMISSION'S RULES
[6 Exam Questions – 6 Groups]

E1A Operating Standards: frequency privileges; emission standards; automatic message forwarding; frequency sharing; stations aboard ships or aircraft

E1B Station restrictions and special operations: restrictions on station location; general operating restrictions, spurious emissions, control operator reimbursement; antenna structure restrictions; RACES operations

E1C Station control: definitions and restrictions pertaining to local, automatic and remote control operation; control operator responsibilities for remote and automatically controlled stations

E1D Amateur Satellite service: definitions and purpose; license requirements for space stations; available frequencies and bands; telecommand and telemetry operations; restrictions, and special provisions; notification requirements

E1E Volunteer examiner program: definitions, qualifications, preparation and administration of exams; accreditation; question pools; documentation requirements

E1F Miscellaneous rules: external RF power amplifiers; national quiet zone; business communications; compensated communications; spread spectrum; auxiliary stations; reciprocal operating privileges; IARP and CEPT licenses; third party communications with foreign countries; special temporary authority

E2 – OPERATING PROCEDURES
[5 Exam Questions – 5 Groups]

E2A Amateur radio in space: amateur satellites; orbital mechanics; frequencies and modes; satellite hardware; satellite operations

E2B Television practices: fast scan television standards and techniques; slow scan television standards and techniques

E2C Operating methods: contest and DX operating; spread-spectrum transmissions; selecting an operating frequency

E2D Operating methods: VHF and UHF digital modes; APRS

E2E Operating methods: operating HF digital modes; error correction

E3 – RADIO WAVE PROPAGATION
[3 Exam Questions – 3 Groups]

E3A Propagation and technique, Earth-Moon-Earth communications; meteor scatter

E3B Propagation and technique, trans-equatorial; long path; gray-line; multi-path propagation

E3C Propagation and technique, Aurora propagation; selective fading; radio-path horizon; take-off angle over flat or sloping terrain; effects of ground on propagation; less common propagation modes

E4 – AMATEUR PRACTICES
[5 Exam Questions – 5 Groups]

E4A Test equipment: analog and digital instruments; spectrum and network analyzers, antenna analyzers; oscilloscopes; testing transistors; RF measurements

E4B Measurement technique and limitations: instrument accuracy and performance limitations; probes; techniques to minimize errors; measurement of "Q"; instrument calibration

E4C Receiver performance characteristics, phase noise, capture effect, noise floor, image rejection, MDS, signal-to-noise-ratio; selectivity

E4D Receiver performance characteristics, blocking dynamic range, intermodulation and cross-modulation interference; 3rd order intercept; desensitization; preselection

E4E Noise suppression: system noise; electrical appliance noise; line noise; locating noise sources; DSP noise reduction; noise blankers

E5 – ELECTRICAL PRINCIPLES
[4 Exam Questions – 4 Groups]

E5A Resonance and Q: characteristics of resonant circuits: series and parallel resonance; Q; half-power bandwidth; phase relationships in reactive circuits

E5B Time constants and phase relationships: RLC time constants: definition; time constants in RL and RC circuits; phase angle between voltage and current; phase angles of series and parallel circuits

E5C Impedance plots and coordinate systems: plotting impedances in polar coordinates; rectangular coordinates

E5D AC and RF energy in real circuits: skin effect; electrostatic and electromagnetic fields; reactive power; power factor; coordinate systems

E6 – CIRCUIT COMPONENTS
[6 Exam Questions – 6 Groups]

E6A Semiconductor materials and devices: semiconductor materials germanium, silicon, P-type, N-type; transistor types: NPN, PNP, junction, field-effect transistors: enhancement mode; depletion mode; MOS; CMOS; N-channel; P-channel

E6B Semiconductor diodes

E6C Integrated circuits: TTL digital integrated circuits; CMOS digital integrated circuits; gates

E6D Optical devices and toroids: cathode-ray tube devices; charge-coupled devices (CCDs); liquid crystal displays (LCDs); toroids: permeability, core material, selecting, winding

E6E Piezoelectric crystals and MMICs: quartz crystals; crystal oscillators and filters; monolithic amplifiers
E6F Optical components and power systems: photoconductive principles and effects, photovoltaic systems, optical couplers, optical sensors, and optoisolators

E7 – PRACTICAL CIRCUITS
[8 Exam Questions – 8 Groups]
E7A Digital circuits: digital circuit principles and logic circuits: classes of logic elements; positive and negative logic; frequency dividers; truth tables
E7B Amplifiers: Class of operation; vacuum tube and solid-state circuits; distortion and intermodulation; spurious and parasitic suppression; microwave amplifiers
E7C Filters and matching networks: filters and impedance matching networks: types of networks; types of filters; filter applications; filter characteristics; impedance matching; DSP filtering
E7D Power supplies and voltage regulators
E7E Modulation and demodulation: reactance, phase and balanced modulators; detectors; mixer stages; DSP modulation and demodulation; software defined radio systems
E7F Frequency markers and counters: frequency divider circuits; frequency marker generators; frequency counters
E7G Active filters and op-amps: active audio filters; characteristics; basic circuit design; operational amplifiers
E7H Oscillators and signal sources: types of oscillators; synthesizers and phase-locked loops; direct digital synthesizers

E8 – SIGNALS & EMISSIONS
[4 Exam Questions – 4 Groups]
E8A AC waveforms: sine, square, sawtooth and irregular waveforms; AC measurements; average and PEP of RF signals; pulse and digital signal waveforms
E8B Modulation and demodulation: modulation methods; modulation index and deviation ratio; pulse modulation; frequency and time division multiplexing

E8C Digital signals: digital communications modes; CW; information rate vs. bandwidth; spread-spectrum communications; modulation methods
E8D Waves, measurements, and RF grounding: peak-to-peak values, polarization; RF grounding

E9 – ANTENNAS & TRANSMISSION LINES
[8 Exam Questions – 8 Groups]
E9A Isotropic and gain antennas: definition; used as a standard for comparison; radiation pattern; basic antenna parameters: radiation resistance and reactance, gain, beamwidth, efficiency
E9B Antenna patterns: E and H plane patterns; gain as a function of pattern; antenna design; Yagi antennas
E9C Wire and phased vertical antennas: beverage antennas; terminated and resonant rhombic antennas; elevation above real ground; ground effects as related to polarization; take-off angles
E9D Directional antennas: gain; satellite antennas; antenna beamwidth; losses; SWR bandwidth; antenna efficiency; shortened and mobile antennas; grounding
E9E Matching: matching antennas to feed lines; power dividers
E9F Transmission lines: characteristics of open and shorted feed lines: 1/8 wavelength; 1/4 wavelength; 1/2 wavelength; feed lines: coax versus open-wire; velocity factor; electrical length; transformation characteristics of line terminated in impedance not equal to characteristic impedance
E9G The Smith chart
E9H Effective radiated power; system gains and losses; radio direction finding antennas

E0 – SAFETY
[1 exam question – 1 group]
E0A Safety: amateur radio safety practices; RF radiation hazards; hazardous materials

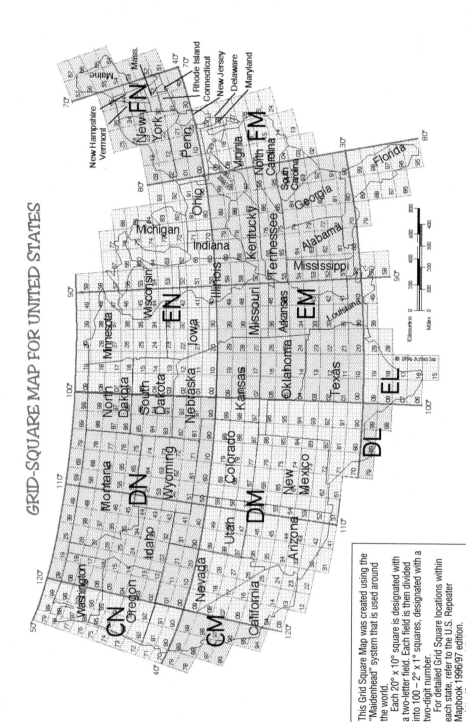

GRID-SQUARE MAP FOR UNITED STATES

This Grid Square Map was created using the "Maidenhead" system that is used around the world.

Each 20° x 10° square is designated with a two-letter field. Each field is then divided into 100 – 2° x 1° squares, designated with a two-digit number.

For detailed Grid Square locations within each state, refer to the U.S. Repeater Mapbook 1996/97 edition.

http://home.earthlink.net/~artsci

© 1996 Artsci Inc

Reprinted by Permission of Artsci Inc.,P.O. Box 1428, Burbank, CA 91507

Appendix

ITU Regions

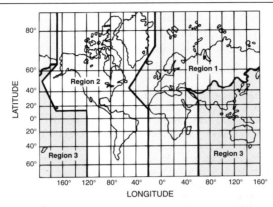

List of Countries Permitting Third-Party Traffic

Country	Call Sign Prefix	Country	Call Sign Prefix	Country	Call Sign Prefix
Antigua/Barbuda	V2	Gambia	C5	Peru	OA-OC
Argentina	LO-LW	Ghana	9G	Philippines	DU-DZ
Australia	VK	Grenada	J3	Pitcairn Island	VR6
Belize	V3	Guatemala	TG	St. Kitts/Nevis	V4
Bolivia	CP	Guyana	8R	St. Lucia	J6
Bosnia-Herzegovina	E7	Haiti	HH	St. Vincent & Grenadines	J8
Brazil	PP-PY	Honduras	HQ-HR	Sierra Leone	9L
Canada	VE, VO, VY	Israel	4X, 4Z	South Africa	ZR-ZU
Chile	CA-CE	Jamaica	6Y	Swaziland	3DA
Colombia	HJ-HK	Jordan	JY	Trinidad/Tobago	9Y-9Z
Comoros	D6	Liberia	EL	Turkey	TA-TC
Costa Rica	TI, TE	Marshall Islands	V7	United Kingdom	GB
Cuba	CM, CO	Mexico	XA-XI	Uruguay	CV-CX
Dominica	J7	Micronesia	V6	Venezuela	YV-YY
Dominican Republic	HI	Nicaragua	YN	ITU – Geneva	4U1ITU
Ecuador	HC-HD	Panama	HO-HP	VIC – Vienna	4U1VIC
El Salvador	YS	Paraguay	ZP		

Countries Holding U.S. Reciprocal Agreements

Antigua, Barbuda	Chile	Greece	Liberia	Seychelles
Argentina	Colombia	Greenland	Luxembourg	Sierra Leone
Australia	Costa Rica	Grenada	Macedonia	Solomon Islands
Austria	Croatia	Guatemala	Marshall Is.	South Africa
Bahamas	Cyprus	Guyana	Mexico	Spain
Barbados	Denmark	Haiti	Micronesia	St. Lucia
Belgium	Dominica	Honduras	Monaco	St. Vincent and
Belize	Dominican Rep.	Iceland	Netherlands	Grenadines
Bolivia	Ecuador	India	Netherlands Ant.	Surinam
Bosnia-	El Salvador	Indonesia	New Zealand	Sweden
Herzegovina	Fiji	Ireland	Nicaragua	Switzerland
Botswana	Finland	Israel	Norway	Thailand
Brazil	France[2]	Italy	Panama	Trinidad, Tobago
Canada[1]	Germany	Jamaica	Paraguay	Turkey
		Japan	Papua New Guinea	Tuvalu
1. Do not need reciprocal permit		Jordan	Peru	United Kingdom[3]
2. Includes all French Territories		Kiribati	Philippines	Uruguay
3. Includes all British Territories		Kuwait	Portugal	Venezuela

POPULAR Q SIGNALS

Given below are a number of Q signals whose meanings most often need to be expressed with brevity and clarity in amateur work. (Q abbreviations take the form of questions only when each is sent followed by a question mark.)

QRG Will you tell me my exact frequency (or that of _____)? Your exact frequency (or that of _____) is _____ kHz.

QRH Does my frequency vary? Your frequency varies.

QRI How is the tone of my transmission? The tone of your transmission is _____ (1. Good; 2. Variable; 3. Bad).

QRJ Are you receiving me badly? I cannot receive you. Your signals are too weak.

QRK What is the intelligibility of my signals (or those of _____)? The intelligibility of your signals (or those of _____) is _____ (1. Bad; 2. Poor; 3. Fair; 4. Good; 5. Excellent).

QRL Are you busy? I am busy (or I am busy with _____). Please do not interfere.

QRM Is my transmission being interfered with? Your transmission is being interfered with _____ (1. Nil; 2. Slightly; 3. Moderately; 4. Severely; 5. Extremely).

QRN Are you troubled by static? I am troubled by static _____ (1-5 as under QRM).

QRO Shall I increase power? Increase power.

QRP Shall I decrease power? Decrease power.

QRQ Shall I send faster? Send faster (_____ WPM).

QRS Shall I send more slowly? Send more slowly (_____WPM).

QRT Shall I stop sending? Stop sending.

QRU Have you anything for me? I have nothing for you.

QRV Are you ready? I am ready.

QRW Shall I inform _____ that you are calling on _____ kHz? Please inform _____ that I am calling on _____ kHz.

QRX When will you call me again? I will call you again at _____ hours (on _____ kHz).

QRY What is my turn? Your turn is numbered _____.

QRZ Who is calling me? You are being called by _____ (on _____ kHz).

QSA What is the strength of my signals (or those of _____)? The strength of your signals (or those of _____) is _____ (1. Scarcely perceptible; 2. Weak; 3. Fairly good; 4. Good; 5. Very good).

QSB Are my signals fading? Your signals are fading.

QSD Is my keying defective? Your keying is defective.

QSG Shall I send _____ messages at a time? Send _____ messages at a time.

QSK Can you hear me between your signals and if so can I break in on your transmission? I can hear you between my signals; break in on my transmission.

QSL Can you acknowledge receipt? I am acknowledging receipt.

QSM Shall I repeat the last message which I sent you, or some previous message? Repeat the last message which you sent me [or message(s) number(s) _____].

QSN Did you hear me (or _____) on _____ kHz? I heard you (or _____) on _____ kHz.

QSO Can you communicate with _____ direct or by relay? I can communicate with _____ direct (or by relay through _____).

QSP Will you relay to _____ ? I will relay to _____.

QST General call preceding a message addressed to all amateurs and ARRL members. This is in effect "CQ ARRL."

QSU Shall I send or reply on this frequency (or on _____ kHz)?

QSW Will you send on this frequency (or on _____ kHz)? I am going to send on this frequency (or on _____ kHz).

QSX Will you listen to _____ on _____ kHz? I am listening to _____ on _____ kHz.

QSY Shall I change to transmission on another frequency? Change to transmission on another frequency (or on _____ kHz).

QSZ Shall I send each word or group more than once? Send each word or group twice (or _____ times).

QTA Shall I cancel message number _____? Cancel message number _____.

QTB Do you agree with my counting of words? I do not agree with your counting of words. I will repeat the first letter or digit of each word or group.

QTC How many messages have you to send? I have messages for you (or for _____).

QTH What is your location? My location is _____.

QTR What is the correct time? The time is _____.

Source: ARRL

ITU EMISSION DESIGNATORS

First Symbol – Modulation system

A	–	Double sideband AM	M	–	Pulse modulated in position/phase

A – Double sideband AM
C – Vestigial sideband AM
D – Amplitude/angle modulated
F – Frequency modulation
G – Phase modulation
H – Single sideband/full carrier
J – Single sideband/suppressed carrier
K – AM pulse
L – Pulse modulated in width/duration

M – Pulse modulated in position/phase
N – Unmodulated carrier
P – Unmodulated pulses
Q – Angle modulated during pulse
R – Single sideband/reduced or variable level carrier
V – Combination of pulse emissions
W – Other types of pulses

Second Symbol – Nature of signal modulating carrier

0 – No modulation
1 – Digital data without modulated subcarrier
2 – Digital data on modulated subcarrier
3 – Analog modulated

7 – Two or more channels of digital data
8 – Two or more channels of analog data
9 – Combination of analog and digital information
X – Other

Third Symbol – Information to be conveyed

A – Manually received telegraphy
B – Automatically received telegraphy
C – Facsimile (FAX)
D – Digital information
E – Voice telephony

F – Video/television
N – No information
W – Combination of these
X – Other

The signal may be "assembled" from the different elements of the emission designators.
For example, an amplitude-modulated voice single sideband signal is J3E:
J 5 single sideband; 3 5 analog, single channel; E 5 telephony.

1. **CW** – International Morse code telegraphy emissions having designators with A, C, H, J or R as the first symbol, 1 as the second symbol, A or B as the third symbol.
2. **DATA** – Telemetry, telecommand and computer communications emissions having designators with A, C, D, F, G, H, J or R as the first symbol: 1 as the second symbol; D as the third symbol; and also emission J2D. Only a digital code of a type specifically authorized in the §Part 97.3 rules may be transmitted.
3. **IMAGE** – Facsimile and television emissions having designators with A, C, D, F, G, H, J or R as the first symbol; 1, 2 or 3 as the second symbol; C or F as the third symbol; and also emissions having B as the first symbol; 7, 8 or 9 as the second symbol; W as the third symbol.
4. **MCW (Modulated carrier wave)** – Tone-modulated international Morse code telegraphy emissions having designators with A, C, D, F, G, H or R as the first symbol; 2 as the second symbol; A or B as the third symbol.
5. **PHONE** – Speech and other sound emissions having designators with A, C, D, F, G, H, J or R as the first symbol; 1, 2 or 3 as the second symbol; E as the third symbol. Also speech emissions having B as the first symbol; 7, 8 or 9 as the second symbol; E as the third symbol. MCW for the purpose of performing the station identification procedure, or for providing telegraphy

practice interspersed with speech, or incidental tones for the purpose of selective calling, or alerting or to control the level of a demodulated signal may also be considered phone.
6. **PULSE** – Emissions having designators with K, L, M, P, Q, V or W as the first symbol; , 1, 2, 3, 7, 8, 9 or X as the second symbol; A, B, C, D, E, F, N, W or X as the third symbol.
7. **RTTY (Radioteletype)** – Narrow-band, direct-printing telegraphy emissions having designators with A, C, D, F, G, H, J or R as the first symbol; 1 as the second symbol; B as the third symbol; and also emission J2B. Only a digital code of a type specifically authorized in the §Part 97.3 rules may be transmitted.
8. **SS (Spread Spectrum)** – Emissions using bandwidth-expansion modulation emissions having designators with A, C, D, F, G, H, J or R as the first symbol; X as the second symbol; X as the third symbol. Only a SS emission of a type specifically authorized in §Part 97.3 rules may be transmitted.
9. **TEST** – Emissions containing no information having the designators with N as the third symbol. Test does not include pulse emissions with no information or modulation unless pulse emissions are also authorized in the frequency band.

Glossary

Advanced: An amateur operator who has passed Element 2, 3A, 3B, and 4A written theory examinations and Element 1A and 1B code tests to demonstrate Morse code proficiency to 13 wpm.

Amateur communication: Non-commercial radio communication by or among amateur stations solely with a personal aim and without personal or business interest.

Amateur operator/primary station license: An instrument of authorization issued by the Federal Communications Commission comprised of a station license, and also incorporating an operator license indicating the class of privileges.

Amateur operator: A person holding a valid license to operate an amateur station issued by the FCC. Amateur operators are frequently referred to as ham operators.

Amateur Radio services: The amateur service, the amateur-satellite service and the radio amateur civil emergency service.

Amateur-satellite service: A radiocommunication service using stations on Earth satellites for the same purpose as those of the amateur service.

Amateur service: A radiocommunication service for the purpose of self-training, intercommunication and technical investigations carried out by amateurs; that is, duly authorized persons interested in radio technique solely with a personal aim and without pecuniary interest.

Amateur station: A station licensed in the amateur service embracing necessary apparatus at a particular location used for amateur communication.

AMSAT: Radio Amateur Satellite Corporation, a non-profit scientific organization.
(850 Sligo Avenue, Silver Spring, MD 20910-4703)

Antenna gain: The increase as a result of the physical construction of the antenna which confines the radiation to desired or useful directions. Usually specified in dB, referenced to the gain of a dipole.

ARES: The emergency division of the American Radio Relay League. See RACES

ARRL: American Radio Relay League, national organization of U.S. Amateur Radio operators. (225 Main Street, Newington, CT 06111)

ATV: Amateur fast-scan television.

Audio Frequency (AF): The range of frequencies that can be heard by the human ear, generally 20 hertz to 20 kilohertz.

Authorized bandwidth: The allowed frequency band, specified in kilohertz, and centered on the carrier frequency.

Automatic control: The use of devices and procedures for station control without the control operator being present at the control point when the station is transmitting.

Automatic Position Reporting System (APRS): Links a remote GPS unit to a ham transceiver which transmits the location to a base station.

Automatic Volume Control (AVC): A circuit that continually maintains a constant audio output volume in spite of deviations in input signal strength.

Beam or Yagi antenna: An antenna array that receives or transmits RF energy in a particular direction. Usually rotatable.

Block diagram: A simplified outline of an electronic system where circuits or components are shown as boxes.

Broadcasting: Information or programming transmitted by means of radio intended for the general public.

Business communications: Any transmission or communication the purpose of which is to facilitate the regular business or commercial affairs of any party. Business communications are prohibited in the amateur service.

Call Book: A published list of all licensed amateur operators available in North American and Foreign editions.

Call sign assignment: The FCC systematically assigns each amateur station their primary call sign.

Carrier frequency: The frequency of an unmodulated electromagnetic wave, usually specified in kilohertz or megahertz.

Certificate of Successful Completion of Examination (CSCE): A certificate verifying successful completion of a written and/or Morse code examination Element. Credit for passing is valid for 365 days.

Coaxial cable, Coax: A concentric, two-conductor cable in which one conductor surrounds the other, separated by an insulator.

Control point: Any place from which a transmitter's function may be controlled.

Control station: A station directly associated with the control point of a radio system, commonly used to control repeaters.

Control operator: An amateur operator designated by the licensee of an amateur station to be responsible for the station transmissions.

Coordinated repeater station: An amateur repeater station for which the transmitting and receiving frequencies have been recommended by the recognized repeater coordinator.

Coordinated Universal Time (UTC): Some-times referred to as Greenwich Mean Time, UCT or Zulu time. The time at the zero-degree (0∞) Meridian which passes through Greenwich, England. A universal time among all amateur operators.

Crystal: A quartz or similar material which has been ground to produce natural vibrations of a specific frequency. Quartz crystals produce a high-degree of frequency stability in radio transmitters.

CW: Continuous wave, another term for the International Morse code.

Dipole antenna: The most common wire antenna. Length is equal to one-half of the wavelength.

Direct Digital Frequency Synthesizer: Modern frequency synthesizer employing technology to reduce phase noise to an insignificant level.

Dummy antenna: A device or resistor which serves as a transmitter's antenna without radiating radio waves. Generally used to tune up a radio transmitter.

Duplex: Transmitting on one frequency, and receiving on another, commonly used in mobile telephone use, as well as mobile business radio on UHF frequencies.

Duplexer: A device that allows a single antenna to be simultaneously used for both reception and transmission.

Effective Radiated Power (ERP): The product of the transmitter (peak envelope) power, expressed in watts, delivered to the antenna, and the relative gain of an antenna over that of a half-wave dipole antenna.

Emergency communication: Any amateur communication directly relating to the immediate safety of life of individuals or the immediate protection of property.

Examination Element: The written theory exams and Morse code test required to obtain amateur radio licenses. Technicians are required to pass written Element 2 theory; Generals are required to pass written Element 3 theory plus Element 1 Morse code test at five (5) words-per-minute; and Extras are required to pass written Element 4 theory.

FCC Form 605: The universal application form used to renew or modify an existing license. (See NCVEC Form 605)

Federal Communications Commission (FCC): A board of five Commissioners, appointed by the President, having the power to regulate wire and radio telecommunications in the United States.

Feedline transmission line: A system of conductors that connects an antenna to a receiver or transmitter.

Field Day: Annual activity sponsored by the ARRL to demonstrate emergency preparedness of amateur operators.

Filter: A device used to block or reduce alternating currents or signals at certain frequencies while allowing others to pass unimpeded.

Frequency: The number of cycles of alternating current in one second.

Frequency coordinator: An individual or organization recognized by amateur operators eligible to engage in repeater operation that recommends frequencies and other operating and/or technical parameters for amateur repeater operation in order to avoid or minimize potential interferences.

Frequency Modulation (FM): A method of varying a radio carrier wave by causing its frequency to vary in accordance with the information to be conveyed.

Frequency privileges: The transmitting frequency bands available to the various classes of amateur operators. The privileges are listed in Part 97.7(a) of the FCC rules.

General: An amateur operator who has passed the Element 3 written theory examination and Element 1 code test to demonstrate Morse code proficiency to 5-wpm.

Ground: A connection, accidental or intentional, between a device or circuit and the earth or some common body and the earth or some common body serving as the earth.

Ground wave: A radio wave that is propagated near or at the earth's surface.

Half-Duplex: A method of operation on a duplex (separate transmit, separate receive) frequency pair where the operator only receives when the microphone button is released. Half-duplex does not permit simultaneous talk and listen.

Handi-Ham system: Amateur organization dedicated to assisting handicapped amateur operators. (3915 Golden Valley Road, Golden Valley, MN 55422)

Harmful interference: Interference which seriously degrades, obstructs or repeatedly interrupts the operation of a radio communication service.

Harmonic: A radio wave that is a multiple of the fundamental frequency. The second harmonic is twice the fundamental frequency, the third harmonic, three times, etc.

Hertz: One complete alternating cycle per second, abbreviated Hz. Named after Heinrich R. Hertz, a German physicist. The number of hertz is the frequency of the audio or radio wave.

High Frequency (HF): The band of frequencies that lie between 3 and 30 megahertz. It is from these frequencies that radio waves are returned to earth from the ionosphere.

High-Pass filter: A device that allows passage of high frequency signals but attenuates the lower frequencies.

Interference: The effect that occurs when two or more radio stations are transmitting at the same time. This includes undesired noise or other radio signals on the same frequency.

International Telecommunication Union (ITU): World organization composed of delegates from member countries, which allocates frequency spectrum for various purposes, including amateur radio.

Ionosphere: Outer limits of atmosphere from which HF amateur communications signals are returned to earth.

Jamming: The intentional, malicious interference with another radio signal.

Key clicks, Chirps: Defective keying of a telegraphy signal sounding like tapping or high varying pitches.

Lid: Amateur slang term for poor radio operator.

Linear amplifier: A device that accurately reproduces a radio wave in magnified form.

Long wire: A horizontal wire antenna that is one wavelength or longer in length.

Low-Pass filter: A device that allows passage of low frequency signals but attenuates the higher frequencies.

Machine: A ham slang word for an automatic repeater station.

Malicious interference: Willful, intentional jamming of radio transmissions.

MARS: The Military Affiliate Radio System. An organization that coordinates the activities of amateur communications with military radio communications.

Maximum authorized transmitting power: Amateur stations must use no more than the maximum transmitter power necessary to carry out the desired communications.

Maximum usable frequency (MUF): The highest frequency that will be returned to earth from the ionosphere.

Medium frequency (MF): The band of frequencies that lies between 300 and 3,000 kHz (3 MHz).

Mobile operation: Radio communications conducted while in motion or during halts at unspecified locations.

Mode: Type of transmission such as voice, teletype, code, television, facsimile.

Modulate: To vary the amplitude, frequency or phase of a radio frequency wave in accordance with the information to be conveyed.

Monolithic Microwave Integrated Circuit (MMIC): A four-legged device designed for RF amplification. Well-suited for VHF, UHF and microwave applications.

Morse code (see CW): The International Morse code, A1A emission. Interrupted continuous wave communications conducted using a dot-dash code for letters, numbers and operating procedure signs.

NCVEC Form 605: The form provided by VECs that is used to apply for an amateur radio license or license upgrade.

Noise level: The strengths of extraneous audible sounds in a given location, usually measured in decibels.

Novice operator: An FCC licensed, entry-level operator in the amateur service. Novices may operate a transmitter in the following meter wavelength bands: 80, 40, 15, 10, 1.25 and 0.23.

Ohm's law: The basic electrical law explaining the relationship between voltage, current and resistance. The current I in a circuit is equal to the voltage E divided by the resistance R, or $I = E/R$.

OSCAR: Acronym for "Orbiting Satellite Carrying Amateur Radio," the name given to a series of satellites designed and built by amateur operators of several nations.

Oscillator: A device for generating oscillations or vibrations of an audio or radio frequency signal.

Output power: The radio-frequency output power of a transmitter's final radio-frequency stage as measured at the output terminal while connected to a load of the impedance recommended by the manufacturer.

Packet radio: A digital method of communicating computer-to-computer. A terminal-node controller makes up the packet of data and directs it to another packet station.

Peak Envelope Power (PEP): 1. The power during one radio frequency cycle at the crest of the modulation envelope, measured under normal operating conditions. 2. The maximum power that can be obtained from a transmitter.

Phone patch: Interconnection of amateur service to the public switched telephone network, and operated by the control operator of the station.

Power supply: A device or circuit that provides the appropriate voltage and current to another device or circuit.

Propagation: The travel of electromagnetic waves or sound waves through a medium.

Q-signals: International three-letter abbreviations beginning with the letter Q used primarily to convey information using the Morse code.

QSL Bureau: An office that bulk processes QSL (radio confirmation) cards for (or from) foreign amateur operators as a postage saving mechanism.

RACES (radio amateur civil emergency service): A radio service using amateur stations for civil defense communications during periods of local, regional, or national civil emergencies.

Radiation: Electromagnetic energy, such as radio waves, traveling forth into space from a transmitter.

Radio Frequency (RF): The range of frequencies over 20 kilohertz that can be propagated through space.

Radio wave: A combination of electric and magnetic fields varying at a radio frequency and traveling through space at the speed of light.

Repeater operation: Automatic amateur stations that retransmit the signals of other amateur stations.

Repeater station: An intermediate station in a system which is arranged to receive a signal from a station, amplify and retransmit the signal to another station. Usually performs this function in both directions, simultaneously.

RST Report: A telegraphy signal report system of Readability, Strength and Tone.

S-meter: A voltmeter calibrated from 0 to 9 that indicates the relative signal strength of an incoming signal at a radio receiver.

Selectivity: The ability of a circuit (or radio receiver) to separate the desired signal from those not wanted.

Sensitivity: The ability of a circuit (or radio receiver) to detect a specified input signal.

Short circuit: An unintended, low-resistance connection across a voltage source resulting in high current and possible damage.

Shortwave: The high frequencies that lie between 3 and 30 megahertz that are propagated long distances.

Simplex: A method of operation of a communication circuit which can receive or transmit, but not both simultaneously. Thus, system stations are operating on the same transmit and receive frequency.

Single-Sideband (SSB): A method of radio transmission in which the RF carrier and one of the sidebands is suppressed and all of the information is carried in the one remaining sideband.

Skip wave, Skip zone: A radio wave reflected back to earth. The distance between the radio transmitter and the site of a radio wave's return to earth.

Sky wave: A radio wave that is refracted back to earth in much the same way that a stone thrown across water skips out. Sometimes called an ionospheric wave.

Spectrum: A series of radiated energies arranged in order of wavelength. The radio spectrum extends from 20 kilohertz upward.

Sporadic-E: Radio propagation caused by refraction of signals by dense ionization at the E-layer of the ionosphere.

Spread spectrum: Modulation method which employs lightning-fast frequency "hopping" to allow multiple users to share a band.

Spurious Emissions: Unwanted radio frequency signals emitted from a transmitter that sometimes causes interference.

Station license, location: No transmitting station shall be operated in the amateur service without being licensed by the FCC. Each amateur station shall have one land location, the address of which appears on the station license.

Sunspot Cycle: An 11-year cycle of solar disturbances which greatly affects radio wave propagation.

TAPR: Tucson Amateur Packet Radio Corporation, a non-profit research and development organization. (8987-309 E. Tangue Verde Rd. #337, Tucson, AZ 85749-9399)

Technician operator: An amateur radio operator who has successfully passed Element 2. This operator has not passed Element 1, the 5-wpm code test.

Technician with Code Credit: An amateur operator who has passed a 5-wpm code test in addition to Technician Class requirements.

Telecommunications: The electrical conversion, switching, transmission and control of audio signals by wire or radio. Also includes video and data communications.

Telegraphy: Telegraphy is communications transmission and reception using CW International Morse Code.

Telephony: Telephony is communications transmission and reception in the voice mode.

Temporary operating authority: Authority to operate your amateur station while awaiting arrival of an upgraded license.

Terrestrial station location: Any location within the major portion of the Earth's atmosphere, including air, sea and land locations.

Third-party traffic: Amateur communication by or under the supervision of the control operator at an amateur station to another amateur station on behalf of others.

Transceiver: A combination radio transmitter and receiver.

Transmatch: An antenna tuner used to match the impedance of the transmitter output to the transmission line of an antenna.

Transmitter: Equipment used to generate radio waves. Most commonly, this radio carrier signal is amplitude or frequency modulated with information and radiated into space.

Transmitter power: The average peak envelope power (output) present at the antenna terminals of the transmitter. The term "transmitted" includes any external radio frequency power amplifier which may be used.

Ultra High Frequency (UHF): Ultra high frequency radio waves that are in the range of 300 to 3,000 MHz.

Upper Sideband (USB): The operating mode for sideband transmissions where the carrier and lower sideband are suppressed.

Vanity call sign: Custom-chosen call sign instead of "next-in-sequence" assignment from the FCC.

Very High Frequency (VHF): Very high frequency radio waves that are in the range of 30 to 300 MHz.

Volunteer Examiner: An amateur operator of at least a General Class level who administers or prepares amateur operator license examinations. A VE must be at least 18 years old and not related to the applicant.

Volunteer Examiner Coordinator (VEC): A member of an organization which has entered into an agreement with the FCC to coordinate the efforts of volunteer examiners in preparing and administering examinations for amateur operator licenses.

Index

LET FORREST M. MIMS, III TEACH YOU ELECTRONICS!

Getting Started in Electronics

This is a complete electronics course in 128 pages! This famous electronics inventor teaches you the basics, takes you on a tour of analog and digital components, explains how they work, and shows how they are combined for various applications. Includes circuit assembly tips and 100 electronic circuits you can build and test. Forrest has written dozens of books, hundreds of articles, invented scientific devices and travelled to the Amazon for NASA, and loves to share his knowledge with eager students! **GSTD $19.95**

ENGINEER'S MINI NOTEBOOKS BY MIMS

These 4 Mini Notebooks include wonderful project ideas for students and hobbyists. Forrest's little books have stimulated many science fair projects and are popular with hobbyists of all ages. Even professional scientists recommend them as a fun way to learn about electronics!

Timer, Op Amp, and Optoelectronic Circuits & Projects This Mini Notebook features more than two dozen 555 timer circuits that you can build, 50 operational amplifier (Op Amp) circuits, and a wide range of optoelectronic circuits and projects including many LED and lightwave circuits. **MINI-1 $12.95**

Science and Communication Circuits & Projects Use these plans to make a simple seismometer; build a sun photo- meter to make accurate measurements of the atmosphere; study rain, lightning, and sunlight; and build a wide variety of lightwave and radio communication circuits. **MINI-2 $12.95**

Electronic Sensor Circuits & Projects Electronic sensors convert light, temperature, sound, and other signals into a form that can be processed by electronic circuits. Learn about solar cells, photoresistors, thermistors, and magnet switches. Then build circuits that respond to heat, pressure, light, and more. **MINI-3 $12.95**

Electronic Formulas, Symbols & Circuits This Mini Notebook provides a complete, basic electronics reference guide for the workshop or your ham shack. Includes many frequently-used formulas, tables, circuit symbols, and device packages. Design and testing tips are provided to help you plan and troubleshoot your circuits. **MINI-4 $12.95**

CUT HERE

FREE
CQ Amateur Radio
Mini-Sub!

Fill in your address information
on the reverse side of this page, fold in half,
tape closed and mail today!

The Radio
Amateur's Journal

25 Newbridge Road
Hicksville, NY 11801-9962

NEED TO FIND A TEST LOCATION FAST?
NEED TO CHANGE YOUR ADDRESS?
NEED TO RENEW YOUR LICENSE?
WANT A VANITY CALL SIGN?

LET THE W5YI-VEC HELP YOU!

The W5YI-VEC is a non-profit organization that coordinates Amateur Radio test sessions throughout the United States and in some foreign countries. Each Volunteer Examiner team is made up of 3 or more licensed Amateur Radio operators certified by The W5YI-VEC to administer ham radio license exams.

The W5YI-VEC can also help you with your FCC Amateur Radio license renewal, change of address, or obtaining a Vanity Call Sign.

COMMERCIAL RADIO LICENSE SERVICES

National Radio Examiners provides examination sites for FCC Commercial Radio Licenses, including GROL, MROP, and the GMDSS license. NRE also can assist you with your license renewal or change of address.

For help with any of these FCC Amateur or Commercial radio licensing services, call or visit:

1-800-669-9594 • www.w5yi.org

BE AN ELMER!

Join the ranks of Gordon West / W5YI amateur radio instructors by helping teach a ham radio class. We provide lots of support to ham instructors to assist them in their efforts to recruit and train new members of the ham radio community. To learn more, visit:

www.haminstructor.com

GIVE AN EXAM!

Another way to help our service and hobby grow is to become a W5YI-VEC Volunteer Examiner. As an Extra Class licensee, you are qualified to help administer all levels of FCC amateur radio license exams. To learn more, or apply, visit:

www.w5yi-vec.org